CAMBRIDGE LIBRARY COLLECTION

Books of enduring scholarly value

Mathematical Sciences

From its pre-historic roots in simple counting to the algorithms powering modern desktop computers, from the genius of Archimedes to the genius of Einstein, advances in mathematical understanding and numerical techniques have been directly responsible for creating the modern world as we know it. This series will provide a library of the most influential publications and writers on mathematics in its broadest sense. As such, it will show not only the deep roots from which modern science and technology have grown, but also the astonishing breadth of application of mathematical techniques in the humanities and social sciences, and in everyday life.

The Collected Mathematical Papers

Arthur Cayley (1821-1895) was a key figure in the creation of modern algebra. He studied mathematics at Cambridge and published three papers while still an undergraduate. He then qualified as a lawyer and published about 250 mathematical papers during his fourteen years at the Bar. In 1863 he took a significant salary cut to become the first Sadleirian Professor of Pure Mathematics at Cambridge, where he continued to publish at a phenomenal rate on nearly every aspect of the subject, his most important work being in matrices, geometry and abstract groups. In 1882 he spent five months at Johns Hopkins University, and in 1883 became president of the British Association for the Advancement of Science. Publication of his Collected Papers - 967 papers in 13 volumes plus an index volume - began in 1889 and was completed after his death under the editorship of his successor in the Sadleirian Chair. This volume contains a complete listing of all the papers, and a thorough index of persons and topics from Abel to Zornow.

The Collected Mathematical Papers

Index

VOLUME 14

ARTHUR CAYLEY

CAMBRIDGE
UNIVERSITY PRESS

CAMBRIDGE UNIVERSITY PRESS

Cambridge New York Melbourne Madrid Cape Town Singapore São Paolo Delhi

Published in the United States of America by Cambridge University Press, New York

www.cambridge.org
Information on this title: www.cambridge.org/9781108005067

© in this compilation Cambridge University Press 2009

This edition first published 1898
This digitally printed version 2009

ISBN 978-1-108-00506-7

MATHEMATICAL PAPERS.

𝔏𝔬𝔫𝔡𝔬𝔫: C. J. CLAY AND SONS,
CAMBRIDGE UNIVERSITY PRESS WAREHOUSE,
AVE MARIA LANE.
𝔊𝔩𝔞𝔰𝔤𝔬𝔴: 263, ARGYLE STREET.

𝔏𝔢𝔦𝔭𝔷𝔦𝔤: F. A. BROCKHAUS.
𝔑𝔢𝔴 𝔜𝔬𝔯𝔨: THE MACMILLAN COMPANY.
𝔅𝔬𝔪𝔟𝔞𝔶: E. SEYMOUR HALE.

THE COLLECTED

MATHEMATICAL PAPERS

OF

ARTHUR CAYLEY, Sc.D., F.R.S.,

LATE SADLERIAN PROFESSOR OF PURE MATHEMATICS IN THE UNIVERSITY OF CAMBRIDGE.

SUPPLEMENTARY VOLUME,

CONTAINING TITLES OF PAPERS AND INDEX.

CAMBRIDGE:

AT THE UNIVERSITY PRESS.

1898.

CAMBRIDGE:

PRINTED BY J. AND C. F. CLAY,
AT THE UNIVERSITY PRESS.

PREFATORY NOTE.

The present volume is supplementary to the series of thirteen volumes which contain the collected mathematical papers of the late Professor Cayley.

The first part is a list of the titles of the papers, extracted from the tables of contents of the respective volumes and arranged in the order in which the papers occur.

The second part is an index of subjects and authors. It has been made by my friend, Mr. F. Howard Collins, who most kindly volunteered to do this laborious work; my expectation is that the index will be a useful guide to the papers.

It is now ten years since the printing of the series of volumes began. For about the first seven of the years, Professor Cayley himself acted as Editor; since his death, the duty has fallen to me. My steady wish has been that unnecessary delay in the publication of the volumes should be avoided; that the wish has been realised, is largely due to the staff of the University Press. Everything that could be done in the way of simplifying my task and assisting its progress has been done with zealous good-will and cordial cooperation. To each and to all of them my thanks are given for the help which has enabled me to fulfil the duty I undertook at the request of the Syndics.

A. R. FORSYTH.

31 *January,* 1898.

COMPLETE LIST

OF

TITLES OF THE PAPERS.

[An Asterisk denotes that the paper is not printed in full.]

VOLUME I.

PAGE

1. *On a Theorem in the Geometry of Position* 1
   Camb. Math. Journ. t. II. (1841), pp. 267—271   (1841)

2. *On the Properties of a certain Symbolical Expression* . . 5
   Camb. Math. Journ. t. III. (1843), pp. 62—71   (1842)

3. *On certain Definite Integrals* 13
   Camb. Math. Journ. t. III. (1843), pp. 138—144   (1842)

4. *On certain Expansions, in series of Multiple Sines and Cosines* 19
   Camb. Math. Journ. t. III. (1843), pp. 162—167   (1842)

5. *On the Intersection of Curves* 25
   Camb. Math. Journ. t. III. (1843), pp. 211—213   (1843)

6. *On the Motion of Rotation of a Solid Body* . . . 28
   Camb. Math. Journ. t. III. (1843), pp. 224—232   (1843)

7. *On a class of Differential Equations, and on the Lines of Curvature of an Ellipsoid* 36
   Camb. Math. Journ. t. III. (1843), pp. 264—267   (1843)

8. *On Lagrange's Theorem* 40
   Camb. Math. Journ. t. III. (1843), pp. 283—286   (1843)

9. *Demonstration of Pascal's Theorem* 43
   Camb. Math. Journ. t. IV. (1845), pp. 18—20   (1843)

10. *On the Theory of Algebraical Curves* 46
    Camb. Math. Journ. t. IV. (1845), pp. 102—112   (1844)

11. *Chapters in the Analytical Geometry of* (n) *Dimensions* . . 55
    Camb. Math. Journ. t. IV. (1845), pp. 119—127   (1844)

12. *On the Theory of Determinants* 63
    Camb. Phil. Trans. t. VIII. (1849), pp. 1—16   (1843)

13. *On the Theory of Linear Transformations* 80
    Camb. Math. Journ. t. IV. (1845), pp. 193—209

14. *On Linear Transformations* 95
    Camb. and Dubl. Math. Journ. t. I. (1846), pp. 104—122

PAGE

15. *Note sur deux Formules données par MM. Eisenstein et Hesse*   113
      Crelle, t. xxix. (1845), pp. 54—57

*16. *Mémoire sur les Hyperdéterminants*   117
      Crelle, t. xxx. (1846), pp. 1—37

*17. *Note on Mr Bronwin's paper on Elliptic Integrals* .   .   .   118
      Camb. Math. Journ. t. iii. (1843), pp. 197, 198

*18. *Remarks on the Rev. B. Bronwin's paper on Jacobi's Theory of*
      Elliptic Functions   119
      Phil. Mag. t. xxii. (1843), pp. 358—368

19. *Investigation of the Transformation of certain Elliptic Functions*   120
      Phil. Mag. t. xxv. (1844), pp. 352—354

20. *On certain results relating to Quaternions*.   .   .   .   123
      Phil. Mag. t. xxvi. (1845), pp. 141—145

*21. *On Jacobi's Elliptic Functions, in reply to the Rev. B. Bronwin:*
      and on Quaternions   127
      Phil. Mag. t. xxvi. (1845), pp. 208—211

22. *On Algebraical Couples*   128
      Phil. Mag. t. xxvii. (1845), pp. 38—40

23. *On the Transformation of Elliptic Functions* .   .   .   132
      Phil. Mag. t. xxvii. (1845), pp. 424—427

24. *On the Inverse Elliptic Functions*   136
      Camb. Math. Journ. t. iv. (1845), pp. 257—277

25. *Mémoire sur les Fonctions doublement périodiques* .   .   156
      Liouville, t. x. (1845), pp. 385—420

26. *Mémoire sur les Courbes du Troisième Ordre* .   .   .   183
      Liouville, t. ix. (1844), pp. 285—293

27. *Nouvelles remarques sur les Courbes du Troisième Ordre*.   .   190
      Liouville, t. x. (1845), pp. 102—109

28. *Sur quelques Intégrales Multiples*   195
      Liouville, t. x. (1845), pp. 158—168

29. *Addition à la Note sur quelques Intégrales Multiples*   .   .   204
      Liouville, t. x. (1845), pp. 242—244

30. *Mémoire sur les Courbes à double Courbure et les Surfaces*
      développables   207
      Liouville, t. x. (1845), pp. 245—250

31. *Démonstration d'un Théorème de M. Chasles* .   .   .   212
      Liouville, t. x. (1845), pp. 383, 384

PAGE

32. *On some Analytical Formulæ and their application to the Theory of Spherical Coordinates* 213
Camb. and Dubl. Math. Journ. t. I. (1846), pp. 22—33

33. *On the Reduction of* $du \div \sqrt{U}$, *when U is a Function of the Fourth Order* 224
Camb. and Dubl. Math. Journ. t. I. (1846), pp. 70—73

34. *Note on the Maxima and Minima of Functions of Three Variables* 228
Camb. and Dubl. Math. Journ. t. I. (1846), pp. 74, 75

35. *On Homogeneous Functions of the Third Order with Three Variables* 230
Camb. and Dubl. Math. Journ. t. I. (1846), pp. 97—104

36. *On the Geometrical Representation of the Motion of a Solid Body* 234
Camb. and Dubl. Math. Journ. t. I. (1846), pp. 164—167

37. *On the Rotation of a Solid Body round a Fixed Point* . . 237
Camb. and Dubl. Math. Journ. t. I. (1846), pp. 167—173 and 264—274

38. *Note on a Geometrical Theorem contained in a Paper by Sir W. Thomson* 253
Camb. and Dubl. Math. Journ. t. I. (1846), pp. 207, 208

39. *On the Diametral Planes of a Surface of the Second Order* . 255
Camb. and Dubl. Math. Journ. t. I. (1846), pp. 274—278

40. *On the Theory of Involution in Geometry* 259
Camb. and Dubl. Math. Journ. t. II. (1847), pp. 52—61

41. *On certain Formulæ for Differentiation, with applications to the evaluation of Definite Integrals* 267
Camb. and Dubl. Math. Journ. t. II. (1847), pp. 122—128

42. *On the Caustic by Reflection at a Circle* 273
Camb. and Dubl. Math. Journ. t. II. (1847), pp. 128—130

43. *On the Differential Equations which occur in Dynamical Problems* 276
Camb. and Dubl. Math. Journ. t. II. (1847), pp. 210—219

44. *On a Multiple Integral connected with the Theory of Attractions* 285
Camb. and Dubl. Math. Journ. t. II. (1847), pp. 219—223

45. *On the Theory of Elliptic Functions* 290
Camb. and Dubl. Math. Journ. t. II. (1847), pp. 256—266

46. *Note on a System of Imaginaries* 301
Phil. Mag. t. xxx. (1847), pp. 257, 258

PAGE

47. *Sur la Surface des Ondes*   302
       Liouville, t. xi. (1846), pp. 291—296

48. *Note sur les Fonctions de M. Sturm* 306
       Liouville, t. xi. (1846), pp. 297—299

49. *Sur quelques Formules du Calcul Intégral*   .   .   . 309
       Liouville, t. xii. (1847), pp. 231—240

50. *Sur quelques Théorèmes de la Géométrie de Position* .   . 317
       Crelle, t. xxxi. (1846), pp. 213—227

51. *Problème de Géométrie Analytique* 329
       Crelle, t. xxxi. (1846), pp. 227—230

52. *Sur quelques Propriétés des Déterminants Gauches* .   . 332
       Crelle, t. xxxii. (1846), pp. 119—123

53. *Recherches sur l'Élimination, et sur la Théorie des Courbes*   . 337
       Crelle, t. xxxiv. (1847), pp. 30—45

54. *Note sur les Hyperdéterminants* 352
       Crelle, t. xxxiv. (1847), pp. 148—152

55. *Sur quelques Théorèmes de la Géométrie de Position* .   . 356
       Crelle, t. xxxiv. (1847), pp. 270—275

56. *Demonstration of a Geometrical Theorem of Jacobi's*   .   . 362
       Camb. and Dubl. Math. Journ. t. iii. (1848), pp. 48, 49

57. *On the Theory of Elliptic Functions* 364
       Camb. and Dubl. Math. Journ. t. iii. (1848), pp. 50, 51

58. *Notes on the Abelian Integrals—Jacobi's System of Differential Equations* 366
       Camb. and Dubl. Math. Journ. t. iii. (1848), pp. 51—54

59. *On the Theory of Elimination* 370
       Camb. and Dubl. Math. Journ. t. iii. (1848), pp. 116—120

60. *On the Expansion of Integral Functions in a series of Laplace's Coefficients* 375
       Camb. and Dubl. Math. Journ. t. iii. (1848), pp. 120, 121

61. *On Geometrical Reciprocity* 377
       Camb. and Dubl. Math. Journ. t. iii. (1848), pp. 173—179

62. *On an Integral Transformation* 383
       Camb. and Dubl. Math. Journ. t. iii. (1848), pp. 286, 287

63. *Démonstration d'un Théorème de M. Boole concernant des Intégrales Multiples* 384
       Liouville, t. xiii. (1848), pp. 245—248

PAGE

64. *Sur la généralisation d'un Théorème de M. Jellett qui se rapporte aux Attractions* 388
Liouville, t. XIII. (1848), pp. 264—268

65. *Nouvelles Recherches sur les Fonctions de M. Sturm* . . 392
Liouville, t. XIII. (1848), pp. 269—274

66. *Sur les Fonctions de Laplace* 397
Liouville, t. XIII. (1848), pp. 275—280

67. *Note sur les Fonctions Elliptiques* 402
Crelle, t. XXXVII. (1848), pp. 58—60

68. *On the application of Quaternions to the Theory of Rotation* . 405
Phil. Mag. t. XXXIII. (1848), pp. 196—200

69. *Sur les Déterminants Gauches* 410
Crelle, t. XXXVIII. (1848), pp. 93—96

70. *Sur quelques Théorèmes de la Géométrie de Position* . . 414
Crelle, t. XXXVIII. (1848), pp. 97—104

71. *Note sur les Fonctions du Second Ordre* 421
Crelle, t. XXXVIII. (1848), pp. 105, 106

72. *Note on the Theory of Permutations* 423
Phil. Mag. t. XXXIV. (1849), pp. 527—529

73. *Abstract of a Memoir by Dr Hesse on the construction of the Surface of the Second Order which passes through nine given points* 425
Camb. and Dubl. Math. Journ. t. IV. (1849), pp. 44—46

74. *On the Simultaneous Transformation of Two Homogeneous Functions of the Second Order* 428
Camb. and Dubl. Math. Journ. t. IV. (1849), pp. 47—50

75. *On the Attraction of an Ellipsoid* 432
Camb. and Dubl. Math. Journ. t. IV. (1849), pp. 50—65

76. *On the Triple Tangent Planes of Surfaces of the Third Order* 445
Camb. and Dubl. Math. Journ. t. IV. (1849), pp. 118—132

77. *On the order of certain Systems of Algebraical Equations* . 457
Camb. and Dubl. Math. Journ. t. IV. (1849), pp. 132—137

78. *Note on the Motion of Rotation of a Solid of Revolution* . 462
Camb. and Dubl. Math. Journ. t. IV. (1849), pp. 268—270

79. *On a System of Equations connected with Malfatti's Problem, and on another Algebraical System* 465
Camb. and Dubl. Math. Journ. t. IV. (1849), pp. 270—275

8   CONTENTS OF VOLUME I.

PAGE

80.   *Sur quelques Transmutations des Lignes Courbes* . . . 471
    Liouville, t. xiv. (1849), pp. 40—46

81.   *Addition au Mémoire sur quelques Transmutations des Lignes Courbes* 476
    Liouville, t. xv. (1850), pp. 351—356

82.   *On the Triadic Arrangements of Seven and Fifteen Things* . 481
    Phil. Mag. t. xxxvii. (1850), pp. 50—53

*83. *On Curves of Double Curvature and Developable Surfaces* . 485
    Camb. and Dubl. Math. Journ. t. v. (1850), pp. 18—22

84.   *On the Developable Surfaces which arise from two Surfaces of the Second Order* 486
    Camb. and Dubl. Math. Journ. t. v. (1850), pp. 46—57

85.   *Note on a Family of Curves of the Fourth Order* . . . 496
    Camb. and Dubl. Math. Journ. t. v. (1850), pp. 148—152

86.   *On the Developable derived from an Equation of the Fifth Order* 500
    Camb. and Dubl. Math. Journ. t. v. (1850), pp. 152—159

*87. *Notes on Elliptic Functions (from Jacobi)* 507
    Camb. and Dubl. Math. Journ. t. v. (1850), pp. 201—204

88.   *On the Transformation of an Elliptic Integral* 508
    Camb. and Dubl. Math. Journ. t. v. (1850), pp. 204—206

89.   *On the Attraction of Ellipsoids (Jacobi's Method)* . . . 511
    Camb. and Dubl. Math. Journ. t. v. (1850), pp. 217—226

90.   *Note sur quelques Formules relatives aux Coniques* . . . 519
    Crelle, t. xxxix. (1850), pp. 1—3

91.   *Sur le Problème des Contacts* 522
    Crelle, t. xxxix. (1850), pp. 4—13

92.   *Note sur un Système de certaines Formules* 532
    Crelle, t. xxxix. (1850), pp. 14, 15

93.   *Note sur quelques Formules qui se rapportent à la Multiplication des Fonctions Elliptiques* 534
    Crelle, t. xxxix. (1850), pp. 16—22

94.   *Note sur l'Addition des Fonctions Elliptiques* 540
    Crelle, t. xli. (1851), pp. 57—65

95.   *Note sur quelques Théorèmes de la Géométrie de Position* . 550
    Crelle, t. xli. (1851), pp. 66—72

96.   *Mémoire sur les Coniques inscrites dans une même Surface du Second Ordre* 557
    Crelle, t. xli. (1851), pp. 73—86

Contents of Volume I.

PAGE

97. *Note sur la Solution de l'Équation* $x^{257} - 1 = 0$ 564
  Crelle, t. XLI. (1851), pp. 81—83

*98. *Note relative à la sixième section du Mémoire sur quelques Théorèmes de la Géométrie de Position* 567
  Crelle, t. XLI. (1851), p. 84

99. *Note sur quelques Formules qui se rapportent à la Multiplication des Fonctions Elliptiques* 568
  Crelle, t. XLI. (1851), pp. 85—92

100. *Note sur la Théorie des Hyperdéterminants* 577
  Crelle, t. XLII. (1851), pp. 368—371

Notes and References to papers in Volume I. 581

Volumes II, III, and IV of the *Cambridge Mathematical Journal* have on the title pages the dates 1841, 1843, 1845 respectively, and volume VIII of the *Cambridge Philosophical Transactions* has the date 1849. As each of these volumes extends over more than a single year, I have added the year of publication for the papers 1, 2, ..., 12. In all other cases, the year of publication is shown by the date on the title page of the volume.

10

VOLUME II.

PAGE

101. *Notes on Lagrange's Theorem* 1
　　　Camb. and Dubl. Math. Journ. t. VI. (1851), pp. 37—45

102. *On a Double Infinite Series.* 8
　　　Camb. and Dubl. Math. Journ. t. VI. (1851), pp. 45—47

103. *On Certain Definite Integrals* 11
　　　Camb. and Dubl. Math. Journ. t. VI. (1851), pp. 136—140

104. *On the Theory of Permutants* 16
　　　Camb. and Dubl. Math. Journ. t. VII. (1852), pp. 40—51

105. *Correction to the Postscript to the Paper on Permutants* . . 27
　　　Camb. and Dubl. Math. Journ. t. VII. (1852), pp. 97, 98

106. *On the Singularities of Surfaces* 28
　　　Camb. and Dubl. Math. Journ. t. VII. (1852), pp. 166—171

107. *On the Theory of Skew Surfaces.* 33
　　　Camb. and Dubl. Math. Journ. t. VII. (1852), pp. 171—173

108. *On certain Multiple Integrals connected with the Theory of Attractions* 35
　　　Camb. and Dubl. Math. Journ. t. VII. (1852), pp. 174—178

109. *On the Rationalisation of certain Algebraical Equations* . . 40
　　　Camb. and Dubl. Math. Journ. t. VIII. (1853), pp. 97—101

110. *Note on the Transformation of a Trigonometrical Expression* . 45
　　　Camb. and Dubl. Math. Journ. t. IX. (1854), pp. 61, 62

111. *On a Theorem of M. Lejeune-Dirichlet's* 47
　　　Camb. and Dubl. Math. Journ. t. IX. (1854), pp. 163—165

112. *Demonstration of a Theorem relating to the Products of Sums of Squares* 49
　　　Phil. Mag. t. IV. (1852), pp. 515—519

113. *On the Geometrical Representation of the Integral*
$$\int dx \div \sqrt{(x+a)(x+b)(x+c)}$$ 53
　　　Phil. Mag. t. v. (1853), pp. 281—284

PAGE

114. *Analytical Researches connected with Steiner's Extension of Malfatti's Problem.* 57
Phil. Trans. t. CXLII. (for 1852), pp. 253—278

115. *Note on the Porism of the In-and-circumscribed Polygon.* . 87
Phil. Mag. t. VI. (1853), pp. 99—102

116. *Correction of two Theorems relating to the In-and-circumscribed Polygon* 91
Phil. Mag. t. VI. (1853), pp. 376, 377

117. *Note on the Integral* $\int dx \div \sqrt{(m-x)(x+a)(x+b)(x+c)}$ . . 93
Phil. Mag. t. VI. (1853), pp. 103—105

118. *On the Harmonic Relation of two Lines or two Points* . . 96
Phil. Mag. t. VI. (1853), pp. 105—107

119. *On a Theorem for the Development of a Factorial* . . . 98
Phil. Mag. t. VI. (1853), pp. 182—185

120. *Note on a Generalisation of the Binomial Theorem* . . . 102
Phil. Mag. t. VI. (1853), p. 185

121. *Note on a Question in the Theory of Probabilities* . . . 103
Phil. Mag. t. VI. (1853), p. 259

122. *On the Homographic Transformation of a Surface of the Second Order into itself* 105
Phil. Mag. t. VI. (1853), pp. 326—333

123. *On the Geometrical Representation of an Abelian Integral* . 113
Phil. Mag. t. VI. (1853), pp. 414—418

124. *On a Property of the Caustic by Refraction of the Circle* . 118
Phil. Mag. t. VI. (1853), pp. 427—431

125. *On the Theory of Groups as depending on the Symbolical Equation* $\theta^n = 1$ 123
Phil. Mag. t. VII. (1854), pp. 40—47

126. *On the Theory of Groups as depending on the Symbolical Equation* $\theta^n = 1$. *Second Part* 131
Phil. Mag. t. VII. (1854), pp. 408, 409

127. *On the Homographic Transformation of a Surface of the Second Order into itself* 133
Phil. Mag. t. VII. (1854), pp. 208—212 : continuation of 122

128. *Developments on the Porism of the In-and-circumscribed Polygon* 138
Phil. Mag. t. VII. (1854), pp. 339—345

  PAGE

129. *On the Porism of the In-and-circumscribed Triangle, and on
      an irrational Transformation of two Ternary Quadratic
      Forms each into itself*   145
          Phil. Mag. t. IX. (1855), pp. 513—517

130. *Deuxième Mémoire sur les Fonctions doublement Périodiques* .   150
          Liouville, t. XIX. (1854), pp. 193—208 : sequel to 25

131. *Nouvelles Recherches sur les Covariants*   164
          Crelle, t. XLVII. (1854), pp. 109—125

132. *Réponse à une Question proposée par M. Steiner* .   .   .   179
          Crelle, t. L. (1855), pp. 277, 278

133. *Sur un Théorème de M. Schläfli*   181
          Crelle, t. L. (1855), pp. 278—282

134. *Remarques sur la Notation des Fonctions Algébriques* .   .   185
          Crelle, t. L. (1855), pp. 282—285

135. *Note sur les Covariants d'une Fonction Quadratique, Cubique,
      ou Biquadratique à deux Indéterminées*   189
          Crelle, t. L. (1855), pp. 285—287

136. *Sur la Transformation d'une Fonction Quadratique en elle-
      même par des Substitutions linéaires*   192
          Crelle, t. L. (1855), pp. 288, 289

137. *Recherches Ultérieures sur les Déterminants gauches* .   .   202
          Crelle, t. L. (1855), pp. 299—313 : continuation of 52 and 69.

138. *Recherches sur les Matrices dont les termes sont des fonctions
      linéaires d'une seule Indéterminée*   216
          Crelle, t. L. (1855), pp. 313—317

139. *An Introductory Memoir on Quantics*   221
          Phil. Trans. t. CXLIV. (for 1854), pp. 244—258

140. *Researches on the Partition of Numbers*   235
          Phil. Trans. t. CXLV. (for 1855), pp. 127—140

141. *A Second Memoir on Quantics*   250
          Phil. Trans. t. CXLVI. (for 1856), pp. 101—126

142. *Numerical Tables Supplementary to Second Memoir on Quantics*   276
          Now first published (1889)

143. *Tables of the Covariants* M *to* W *of the Binary Quintic : from
      the Second, Third, Fifth, Eighth, Ninth and Tenth Memoirs
      on Quantics*   282
          Arranged in the present form (1889)

CONTENTS OF VOLUME II.   13

PAGE

144. *A Third Memoir on Quantics* 310
       Phil. Trans. t. CXLVI. (for 1856), pp. 627—647

145. *A Memoir on Caustics* 336
       Phil. Trans. t. CXLVII. (for 1857), pp. 273—312

146. *A Memoir on Curves of the Third Order.* 381
       Phil. Trans. t. CXLVII. (for 1857), pp. 415—446

147. *A Memoir on the Symmetric Functions of the Roots of an Equation* 417
       Phil. Trans. t. CXLVII. (for 1857), pp. 489—496

148. *A Memoir on the Resultant of a System of two Equations* . 440
       Phil. Trans. t. CXLVII. (for 1857), pp. 703—715

149. *On the Symmetric Functions of the Roots of certain Systems of two Equations* 454
       Phil. Trans. t. CXLVII. (for 1857), pp. 717—726

150. *A Memoir on the Conditions for the Existence of given Systems of Equalities among the Roots of an Equation* . . . 465
       Phil. Trans. t. CXLVII. (for 1857), pp. 727—731

151. *Tables of the Sturmian Functions for Equations of the Second, Third, Fourth, and Fifth Degrees* 471
       Phil. Trans. t. CXLVII. (for 1857), pp. 733—736

152. *A Memoir on the Theory of Matrices* 475
       Phil. Trans. t. CXLVIII. (for 1858), pp. 17—37

153. *A Memoir on the Automorphic Linear Transformation of a Bipartite Quadric Function* 497
       Phil. Trans. t. CXLVIII. (for 1858), pp. 39—46

154. *Supplementary Researches on the Partition of Numbers* . . 506
       Phil. Trans. t. CXLVIII. (for 1858), pp. 47—52

155. *A Fourth Memoir on Quantics* 513
       Phil. Trans. t. CXLVIII. (for 1858), pp. 415—427

156. *A Fifth Memoir on Quantics* 527
       Phil. Trans. t. CXLVIII. (for 1858), pp. 429—460

157. *On the Tangential of a Cubic* 558
       Phil. Trans. t. CXLVIII. (for 1858), pp. 461—463

158. *A Sixth Memoir on Quantics* 561
       Phil. Trans. t. CXLIX. (for 1859), pp. 61—90

Notes and References to papers in Volume II. 593

14

VOLUME III.

PAGE

159. *On some Integral Transformations* 1
       Quart. Math. Journ. t. I. (1857), pp. 4—6

160. *On a Theorem relating to Reciprocal Triangles* . . . 5
       Quart. Math. Journ. t. I. (1857), pp. 7—10

161. *A problem in Permutations* 8
       Quart. Math. Journ. t. I. (1857), p. 79

162. *Two letters on Cubic Forms* 9
       Quart. Math. Journ. t. I. (1857), pp. 85—87 and 90, 91

163. *On Hansen's Lunar Theory* 13
       Quart. Math. Journ. t. I. (1857), pp. 112—125

164. *On Gauss' Theory for the Attraction of Ellipsoids* . . . 25
       Quart. Math. Journ. t. I. (1857), pp. 162—166

165. *On some Geometrical Theorems relating to a triangle circum-*
     scribed about a Conic 29
       Quart. Math. Journ. t. I. (1857), pp. 169—175

166. *Note on the Homology of Sets* 35
       Quart. Math. Journ. t. I. (1857), p. 178

167. *Apropos of Partitions* 36
       Quart. Math. Journ. t. I. (1857), pp. 183, 184

*168. *A demonstration of the Fundamental Property of Geodesic*
     Lines 38
       Quart. Math. Journ. t. I. (1857), pp. 185, 186

169. *Eisenstein's Geometrical Proof of the Fundamental Theorem for*
     Quadratic Residues (*Translated from the Original Memoir,*
     Crelle, t. XXVIII. (1844), *with an addition by A. Cayley*) . 39
       Quart. Math. Journ. t. I. (1857), pp. 186—191

170. *On Schellbach's Solution of Malfatti's Problem* 44
       Quart. Math. Journ. t. I. (1857), pp. 222—226

PAGE

171. *Note on Mr Salmon's Equation of the Orthotomic Circle* . 48
   Quart. Math. Journ. t. I. (1857), pp. 242—244

172. *Note on the Logic of Characteristics* 51
   Quart. Math. Journ. t. I. (1857), pp. 257—259

173. *On Laplace's Method for the Attraction of Ellipsoids* . . 53
   Quart. Math. Journ. t. I. (1857), pp. 285—300

*174. *On the Oval of Descartes* 66
   Quart. Math. Journ. t. I. (1857), pp. 320—328

175. *On the Porism of the In-and-circumscribed Triangle* . . 67
   Quart. Math. Journ. t. I. (1857), pp. 344—354

176. *Note on Jacobi's Canonical Formulæ for Disturbed Motion in*
   an Elliptic Orbit 76
   Quart. Math. Journ. t. I. (1857), pp. 355, 356

177. *Solution of a Mechanical Problem* 78
   Quart. Math. Journ. t. I. (1857), pp. 405, 406

178. *On the à posteriori Demonstration of the Porism of the In-and-*
   circumscribed Triangle 80
   Quart. Math. Journ. t. II. (1858), pp. 31—38

179. *On certain Forms of the Equation of a Conic* . . . 86
   Quart. Math. Journ. t. II. (1858), pp. 44—48

180. *Note on the Reduction of an Elliptic Orbit to a fixed plane* . 91
   Quart. Math. Journ. t. II. (1858), pp. 49—54

181. *On Sir W. R. Hamilton's Method for the Problem of three*
   or more Bodies 97
   Quart. Math. Journ. t. II. (1858), pp. 66—73

182. *On Lagrange's Solution of the Problem of two fixed Centres* . 104
   Quart. Math. Journ. t. II. (1858), pp. 76—83

183. *Note on Certain Systems of Circles* 111
   Quart. Math. Journ. t. II. (1858), pp. 83—88

184. *A Theorem relating to Surfaces of the Second Order* . . 115
   Quart. Math. Journ. t. II. (1858), pp. 140—142

185. *Note on the ' Circular Relation' of Prof. Möbius* . . . 118
   Quart. Math. Journ. t. II. (1858), p. 162

186. *On the determination of the value of a certain Determinant* . 120
   Quart. Math. Journ. t. II. (1858), pp. 163—166

187. *On the Sums of Certain Series arising from the Equation*
   $x = u + tfx$ 124
   Quart. Math. Journ. t. II. (1858), pp. 167—171

PAGE

188. *On the Simultaneous Transformation of two Homogeneous Functions of the Second Order* 129
    Quart. Math. Journ. t. II. (1858), pp. 192—195

189. *Note on a formula in finite Differences* 132
    Quart. Math. Journ. t. II. (1858), pp. 198—201

190. *On the System of Conics which pass through the same four points* 136
    Quart. Math. Journ. t. II. (1858), pp. 206, 207

191. *Note on the Expansion of the true Anomaly* 139
    Quart. Math. Journ. t. II. (1858), pp. 229—232

192. *On the Area of the Conic Section represented by the General Trilinear Equation of the Second Degree* 143
    Quart. Math. Journ. t. II. (1858), pp. 248—253

193. *On Rodrigues' Method for the Attraction of Ellipsoids* . . 149
    Quart. Math. Journ. t. II. (1858), pp. 333—337

194. *Note on the Theory of Attraction* 154
    Quart. Math. Journ. t. II. (1858), pp. 338, 339

195. *Report on the Recent Progress of Theoretical Dynamics* . 156
    Report of British Association, 1857, pp. 1—42

196. *Note sur un Problème d'Analyse Indéterminée* . . . 205
    Nouvelles Annales de Math. t. XVI. (1857), pp. 161—165

197. *Note on the Theory of Logarithms* 208
    Phil. Mag. t. XI. (1856), pp. 275—280

198. *Note on a Result of Elimination* 214
    Phil. Mag. t. XI. (1856), pp. 378, 379

199. *Note on the Theory of Elliptic Motion* 216
    Phil. Mag. t. XI. (1856), pp. 425—428

200. *On the Cones which pass through a given Curve of the Third Order in Space* 219
    Phil. Mag. t. XII. (1856), pp. 20—22

201. *Second Note on the Theory of Logarithms* 222
    Phil. Mag. t. XII. (1856), pp. 354—360

202. *Supplementary Remarks on the Porism of the In-and-circumscribed Triangle* 229
    Phil. Mag. t. XIII. (1857), pp. 19—30

203. *On the Theory of the Analytical Forms called Trees* . . 242
    Phil. Mag. t. XIII. (1857), pp. 172—176

PAGE

204. *On a Problem in the Partition of Numbers* 247
Phil. Mag. t. XIII. (1857), pp. 245—248

205. *Note on the Summation of a Certain Factorial Expression* . 250
Phil. Mag. t. XIII. (1857), pp. 419—423

206. *Note on a Theorem relating to the Rectangular Hyperbola* . 254
Phil. Mag. t. XIII. (1857), p. 423

207. *Analytical Solution of the Problem of Tactions* . . . 255
Phil. Mag. t. XIII. (1857), pp. 507—509

208. *Note on the Equipotential Curve $\dfrac{m}{r} + \dfrac{m'}{r'} = C$* 258
Phil. Mag. t. XIV. (1857), pp. 142—146

209. *A Demonstration of Sir W. R. Hamilton's Theorem of the Isochronism of the Circular Hodograph* 262
Phil. Mag. t. XIV. (1857), pp. 427—430

210. *On the Cubic Transformation of an Elliptic Function* . . 266
Phil. Mag. t. XV. (1858), pp. 363, 364

211. *On a Theorem relating to Hypergeometric Series* . . . 268
Phil. Mag. t. XVI. (1858), pp. 356, 357

212. *A Memoir on the Problem of Disturbed Elliptic Motion* . 270
Mem. R. Astron. Soc. t. XXVII. (1859), pp. 1—29

213. *On the Development of the Disturbing Function in the Lunar Theory* 293
Mem. R. Astron. Soc. t. XXVII. (1859), pp. 69—95

214. *The First part of a Memoir on the Development of the Disturbing Function in the Lunar and Planetary Theories* . 319
Mem. R. Astron. Soc. t. XXVIII. (1860), pp. 187—215

215. *A Supplementary Memoir on the Problem of Disturbed Elliptic Motion* 344
Mem. R. Astron. Soc. t. XXVIII. (1860), pp. 217—234

216. *Tables of the Development of Functions in the Theory of Elliptic Motion* 360
Mem. R. Astron. Soc. t. XXIX. (1861), pp. 191—306

217. *A Memoir on the Problem of the Rotation of a solid Body* . 475
Mem. R. Astron. Soc. t. XXIX. (1861), pp. 307—342

218. *A Third Memoir on the Problem of Disturbed Elliptic Motion* 505
Mem. R. Astron. Soc. t. XXXI. (1863), pp. 43—56

PAGE

219. *On some formulæ relating to the Variation of the Plane of a*
      Planet's Orbit   516
              Monthly Notices R. Astron. Soc. t. XXI. (1861), pp. 43—46

220. *Note on a Theorem of Jacobi's in relation to the Problem of*
      Three Bodies   519
              Monthly Notices R. Astron. Soc. t. XXII. (1862), pp. 76—78

221. *On the Secular Acceleration of the Moon's Mean Motion*   .   522
              Monthly Notices R. Astron. Soc. t. XXII. (1862), pp. 173—230

222. *On Lambert's Theorem for Elliptic Motion*   562
              Monthly Notices R. Astron. Soc. t. XXII. (1862), pp. 238—242

Notes and References to papers in Volume III..   567

VOLUME IV.

PAGE

223. *Note sur un théorème général par rapport à l'élimination* . 1
      Tortolini, t. VII. (1856), pp. 454—458

224. *Sur un théorème d'Abel. Note* 5
      Tortolini, t. VIII. (1857), pp. 201—203

225. *On a Class of Dynamical Problems* 7
      Proc. R. S. t. VIII. (1857), pp. 506—511

226. *On Professor Mac Cullagh's Theorem of the Polar Plane* . . 12
      Proc. R. I. A. t. VI. (1858), pp. 481—491

227. *On the Theory of Reciprocal Surfaces* 21
      Proc. R. I. A. t. VII. (1862), pp. 20—28

228. *Sur l'intégrale* $\int_0^1 \dfrac{t^{\mu+\frac{1}{2}}(1-t)^{\mu-\frac{1}{2}}\,dt}{(a+bt-ct^2)^{\mu+1}}$ 28
      Liouville, t. II. (1857), pp. 49—51

229. *Note sur une formule pour la réversion des séries* . . . 30
      Crelle, t. LII. (1856), pp. 276—284

230. *Note sur la méthode d'élimination de Bezout* 38
      Crelle, t. LIII. (1857), pp. 366, 367

231. *Note sur l'équation* $x^2 - Dy^2 = \pm 4$, $D \equiv 5 \,(\mathrm{mod.}\ 8)$ . . . 40
      Crelle, t. LIII. (1857), pp. 369—371

232. *Mémoire sur la forme canonique des fonctions binaires* . . 43
      Crelle, t. LIV. (1857), pp. 48—58

233. *Addition au Mémoire sur la forme canonique des fonctions binaires* 53
      Crelle, t. LIV. (1857), p. 292

234. *Deuxième Note sur une formule pour la réversion des séries* . 54
      Crelle, t. LIV. (1857), pp. 156—161

235. *Sur quelques formules pour la transformation des intégrales elliptiques* 60
      Crelle, t. LV. (1858), pp. 15—24

PAGE

236. *Note sur la composition du nombre 47 par rapport aux vingt-troisièmes racines de l'unité*   70
        Crelle, t. LV. (1858), p. 192

237. *Théorème sur les déterminants gauches*   72
        Crelle, t. LV. (1858), pp. 277, 278

238. *Note sur les normales d'une conique*   74
        Crelle, t. LVI. (1859), pp. 182—185

239. *Addition à la Note sur la composition du nombre 47 par rapport aux vingt-troisièmes racines de l'unité* . . .   78
        Crelle, t. LVI. (1859), pp. 186, 187

240. *Note on a Theorem in Spherical Trigonometry* . . .   80
        Phil. Mag. t. XVII. (1859), p. 151

241. *On Poinsot's four new Regular Solids*   81
        Phil. Mag. t. XVII. (1859), pp. 123—128

242. *Second Note on Poinsot's four new Regular Polyhedra* . .   86
        Phil. Mag. t. XVII. (1859), pp. 209, 210

243. *On the Theory of Groups as depending on the Symbolic Equation $\theta^n = 1$. Third Part*   88
        Phil. Mag. t. XVIII. (1859), pp. 34—37. Sequel to 125 and 126

244. *On an Analytical Theorem relating to the distribution of Electricity upon two Spherical Surfaces*   92
        Phil. Mag. t. XVIII. (1859), pp. 119—127

245. *On an Analytical Theorem connected with the distribution of Electricity upon two Spherical Surfaces. Second Part* .   99
        Phil. Mag. t. XVIII. (1859), pp. 193—202

246. *On Contour and Slope Lines* 108
        Phil. Mag. t. XVIII. (1859), pp. 264—268

247. *On the Analytical Forms called Trees. Second Part* . . 112
        Phil. Mag. t. XVIII. (1859), pp. 374—378. Continuation of 203

248. *Sketch of a proof of the Theorem that every Algebraic Equation has a Root* 116
        Phil. Mag. t. XVIII. (1859), pp. 436—439

249. *Note on Cones of the Third Order* 120
        Phil. Mag. t. XVIII. (1859), pp. 439—442

250. *Sur la surface qui est l'enveloppe des plans conduits par les points d'un ellipsoïde perpendiculairement aux rayons menés par le centre* 123
        Tortolini, t. II. (1859), pp. 168—179

PAGE

251. *Sur quelques formules pour la différentiation*   135
      Tortolini, t. II. (1859), pp. 214—230

252. *Note sur l'équation des différences pour une équation donnée
      de degré quelconque*   150
      Tortolini, t. II. (1859), pp. 365, 366

253. *Sur la courbe parallèle à l'ellipse*   152
      Tortolini, t. III. (1860), pp. 311—316

254. *Sur la surface parallèle à l'ellipsoïde*   158
      Tortolini, t. III. (1860), pp. 345—352

255. *On a Problem of Double Partitions.*   166
      Phil. Mag. t. XX. (1860), pp. 337—341

256. *On a System of Algebraic Equations*   171
      Phil. Mag. t. XX. (1860), pp. 341, 342

257. *On the Cubic Centres of a line with respect to three lines and
      a line*   173
      Phil. Mag. t. XX. (1860), pp. 418—423

258. *On a Relation between two Ternary Cubic Forms* . . .   179
      Phil. Mag. t. XX. (1860), pp. 512—514

259. *The Problem of Polyhedra*   182
      Phil. Trans. t. CXLVII. (for 1857), pp. 183—185

260. *On the Double Tangents of a Plane Curve*   186
      Phil. Trans. t. CXLIX. (for 1859), pp. 193—212

261. *On the Conic of Five-pointic Contact at any Point of a Plane
      Curve*   207
      Phil. Trans. t. CXLIX. (for 1859), pp. 371—400

262. *On the Equation of Differences for an Equation of any Order
      and in particular for the Equations of the Orders two,
      three, four, and five*   240
      Phil. Trans. t. CL. (for 1860), pp. 93—112

263. *Demonstration of a Theorem in Finite Differences* . . .   262
      Phil. Trans. t. CL. (for 1860), pp. 321—323

264. *On an Extension of Arbogast's Method of Derivations* . .   265
      Phil. Trans. t. CLI. (for 1861), pp. 37—43

265. *Addition to the Memoir on an Extension of Arbogast's Method
      of Derivations*   272
      Now first published, 1891

PAGE

266. *On the Equation for the Product of the Differences of all but one of the Roots of a given Equation* 276
Phil. Trans. t. CLI. (for 1861), pp. 45—59

267. *On the Porism of the In-and-circumscribed Polygon* . . 292
Phil. Trans. t. CLI. (for 1861), pp. 225—239

268. *On a New Auxiliary Equation in the Theory of Equations of the Fifth Order* 309
Phil. Trans. t. CLI. (for 1861), pp. 263—276

269. *A Seventh Memoir on Quantics* 325
Phil. Trans. t. CLI. (for 1861), pp. 277—292

270. *On the Double Tangents of a Curve of the Fourth Order* . 342
Phil. Trans. t. CLI. (for 1861), pp. 357—362

271. *Sur l'invariant le plus simple d'une fonction quadratique bi-ternaire, et sur le résultant de trois fonctions quadratiques ternaires* 349
Crelle, t. LVII. (1860), pp. 139—148

272. *Démonstration d'un théorème de Jacobi par rapport au problème de Pfaff* 359
Crelle, t. LVII. (1860), pp. 273—277

273. *Note sur la transformation de Tschirnhausen* 364
Crelle, t. LVIII. (1861), pp. 259—262

274. *Deuxième Note sur la transformation de Tschirnhausen* . . 368
Crelle, t. LVIII. (1861), pp. 263—269

275. *On Tschirnhausen's Transformation* 375
Phil. Trans. t. CLI. (1861), pp. 561—578

276. *On the Analytical Theory of the Conic* 395
Phil. Trans. t. CLII. (1862), pp. 639—662

277. *On the Wave Surface* 420
Quart. Math. Journ. t. III. (1860), pp. 16—22

278. *Note on the Singular Solutions of Differential Equations* . 426
Quart. Math. Journ. t. III. (1860), pp. 36, 37

279. *On a Theorem relating to Spherical Conics* 428
Quart. Math. Journ. t. III. (1860), p. 53

280. *On the Conics which touch four given lines* 429
Quart. Math. Journ. t. III. (1860), pp. 94—96

281. *Note on the Wave Surface* 432
Quart. Math. Journ. t. III. (1860), pp. 142—144

PAGE

282. *On a particular case of Castillon's Problem* 435
Quart. Math. Journ. t. III. (1860), pp. 157—164

283. *On a Theorem relating to Homographic Figures* . . . 442
Quart. Math. Journ. t. III. (1860), pp. 177—180

284. *On a New Analytical Representation of Curves in Space* . 446
Quart. Math. Journ. t. III. (1860), pp. 225—236

285. *On the System of Conics having double contact with each other* 456
Quart. Math. Journ. t. III. (1860), pp. 246—250

286. *Note on the value of certain Determinants the terms of which
are the squared distances of Points in a Plane or in Space* 460
Quart. Math. Journ. t. III. (1860), pp. 275—277

287. *Note on the Equation for the squared differences of the Roots
of a Cubic Equation* 463
Quart. Math. Journ. t. III. (1860), pp. 307—309

288. *Note on the Curvature of a plane Curve at a double point,
and on the Curvature of Surfaces* 466
Quart. Math. Journ. t. III. (1860), pp. 322—326

289. *On some Numerical Expansions* 470
Quart. Math. Journ. t. III. (1860), pp. 366—369

290. *A Discussion of the Sturmian Constants for cubic and quartic
Equations* 473
Quart. Math. Journ. t. IV. (1861), pp. 7—12

291. *On the Demonstration of a Theorem relating to the Moments
of Inertia of a Solid Body* 478
Quart. Math. Journ. t. IV. (1861), pp. 25—27

292. *A Theorem in Conics* 481
Quart. Math. Journ. t. IV. (1861), pp. 131—133

293. *On a Certain System of Functional Symbols* 484
Quart. Math. Journ. t. IV. (1861), pp. 225—230

294. *On a New Analytical Representation of Curves in Space* . 490
Quart. Math. Journ. t. V. (1862), pp. 81—86

295. *On the Construction of the ninth point of Intersection of the
Cubics which pass through eight given points* . . . 495
Quart. Math. Journ. t. V. (1862), pp. 222—233

296. *On the Conics which pass through the four foci of a given
Conic* 505
Quart. Math. Journ. t. V. (1862), pp. 275—280

PAGE

297.   *On some Formulæ relating to the Distances of a point from the vertices of a Triangle and to the Problem of Tactions.*   510
     Quart. Math. Journ. t. v. (1862), pp. 381—384

298.   *Report on the Progress of the Solution of certain Special Problems of Dynamics*   513
     Report of the Brit. Assoc. for the Advancement of Science, 1862, pp. 184—252

299.   *Mathematics, recent Terminology in*   594
     English Cyclopædia, t. v. (1860), pp. 534—542

Notes and References to papers in Volume IV.   609

VOLUME V.

PAGE

300. *Note relative aux droites en involution de M. Sylvester* . . 1
      Comptes Rendus, Paris, t. LII. (1861), pp. 1039—1042

301. *Sur les cônes du second ordre qui passent par six points donnés* 4
      Comptes Rendus, Paris, t. LII. (1861), pp. 1216—1218

302. *Considérations générales sur les courbes en espace* . . . 7
      Comptes Rendus, Paris, t. LIV. (1862), pp. 55—60, 396—400, 672—678

303. *Sur le problème du polygone inscrit et circonscrit. Lettre à M. Poncelet* 21
      Comptes Rendus, Paris, t. LV. (1862), pp. 700, 701

304. *Sur un mémoire de Jacobi. Extrait d'une lettre à M. J. Bertrand* 23
      Comptes Rendus, Paris, t. LVI. (1863), p. 43

305. *Considérations générales sur les courbes en espace. Courbes du cinquième ordre* 24
      Comptes Rendus, Paris, t. LVIII. (1864), pp. 994—1000

306. *Sur les coniques qui touchent des courbes d'ordre quelconque. Extrait d'une lettre à M. Chasles* 31
      Comptes Rendus, Paris, t. LIX. (1864), pp. 224, 225

307. *Note sur les fonctions* al (x), *&c., de M. Weierstrass* . . 33
      Liouville, t. VII. (1862), pp. 137—142

308. *On the △ faced Polyacrons, in reference to the problem of the enumeration of Polyhedra* 38
      Manchester Memoirs, t. I. (1862), pp. 248—256

309. *Note on the Theory of Determinants* 45
      Phil. Mag. t. XXI. (1861), pp. 180—185

310. *Note on Mr Jerrard's researches on the Equation of the Fifth Order* 50
      Phil. Mag. t. XXI. (1861), pp. 210—214

PAGE

311.   *On a Theorem of Abel's relating to Equations of the Fifth Order*   55
        Phil. Mag. t. XXI. (1861), pp. 257—263

312.   *On the Partitions of a Close*   62
        Phil. Mag. t. XXI. (1861), pp. 424—428

313.   *On a Surface of the Fourth Order*   66
        Phil. Mag. t. XXI. (1861), pp. 491—495

314.   *On the Curves situate on a Surface of the Second Order* .   70
        Phil. Mag. t. XXII. (1861), pp. 35—38

315.   *On the Cubic Centres of a Line with respect to three lines and a line*   73
        Phil. Mag. t. XXII. (1861), pp. 433—436

*316. *Note on the solution of an Equation of the Fifth Order* . .   77
        Phil. Mag. t. XXIII. (1862), pp. 195, 196

317.   *Note on the transformation of a certain Differential Equation*   78
        Phil. Mag. t. XXIII. (1862), pp. 266, 267

*318. *On a question in the Theory of Probabilities*   80
        Phil. Mag. t. XXIII. (1862), pp. 361—365

*319. *Postscript to the paper, On a question in the Theory of Probabilities*   85
        Phil. Mag. t. XXIII. (1862), pp. 470, 471

320.   *On the Transcendent* $\mathrm{gd}\, u = \frac{1}{i} \log \tan \left(\frac{1}{4}\pi + \frac{1}{2}ui\right)$ 86
        Phil. Mag. t. XXIV. (1862), pp. 19—21

*321. *Final Remarks on Mr Jerrard's theory of Equations of the Fifth Order*   89
        Phil. Mag. t. XXIV. (1862), p. 290

322.   *On a Skew Surface of the Third Order*   90
        Phil. Mag. t. XXIV. (1862), pp. 514—519

323.   *On a tactical Theorem relating to the Triads of Fifteen Things*   95
        Phil. Mag. t. XXV. (1863), pp. 59—61

324.   *Note on a Theorem relating to Surfaces*   98
        Phil. Mag. t. XXV. (1863), pp. 61, 62

325.   *Note on a Theorem relating to a Triangle, Line, and Conic* .   100
        Phil. Mag. t. XXV. (1863), pp. 181—183

PAGE

326. *Theorems relating to the Canonic Roots of a Binary Quantic of an odd order* 103
Phil. Mag. t. xxv. (1863), pp. 206—208

327. *On the Stereographic Projection of the Spherical Conic* . . 106
Phil. Mag. t. xxv. (1863), pp. 350—353

328. *On the delineation of a Cubic Scroll* 110
Phil. Mag. t. xxv. (1863), pp. 528—530

329. *Note on the Problem of Pedal Curves* 113
Phil. Mag. t. xxvi. (1863), pp. 20, 21

330. *On Differential Equations and Umbilici* 115
Phil. Mag. t. xxvi. (1863), pp. 373—379 and 441—452

331. *Analytical Theorem relating to the four Conics inscribed in the same Conic and passing through the same three Points* 131
Phil. Mag. t. xxvii. (1864), pp. 42, 43

332. *Analytical Theorem relating to the sections of a Quadric Surface* 133
Phil. Mag. t. xxvii. (1864), pp. 43, 44

333. *Note on the Nodal Curve of the Developable derived from the Quartic Equation $(a, b, c, d, e\!\!\Sigma t, 1)^4 = 0$* . . . 135
Phil. Mag. t. xxvii. (1864), pp. 437—440

334. *Note on the Theory of Cubic Surfaces* 138
Phil. Mag. t. xxvii. (1864), pp. 493—496

335. *Tables des formes quadratiques binaires pour les déterminants négatifs depuis $D = -1$ jusqu'à $D = -100$, pour les déterminants positifs non carrés depuis $D = 2$ jusqu'à $D = 99$, et pour les treize déterminants négatifs irréguliers qui se trouvent dans le premier millier* 141
Crelle, t. lx. (1862), pp. 357—372

336. *Note sur l'élimination* 157
Crelle, t. lx. (1862), pp. 373, 374

337. *Note sur la réalité des racines d'une équation quadratique* . 160
Crelle, t. lxi. (1863), pp. 367, 368

338. *Nouvelles recherches sur l'élimination et la théorie des courbes* 162
Crelle, t. lxiii. (1864), pp. 34—39

339. *On Skew Surfaces, otherwise Scrolls.* 168
Phil. Trans. t. cliii. (for 1863), pp. 453—483

340. *A Second Memoir on Skew Surfaces, otherwise Scrolls* . . 201
Phil. Trans. t. cliv. (for 1864), pp. 559—576

28 CONTENTS OF VOLUME V.

PAGE

341. *On the Sextactic Points of a Plane Curve* 221
Phil. Trans. t. CLV. (for 1865), pp. 545—578

342. *On the Conics which pass through three given points and touch a given line* 258
Quart. Math. Journ. t. VI. (1864), pp. 24—30

343. *On the Cusp of the second kind or Nodecusp.* . . . 265
Quart. Math. Journ. t. VI. (1864), pp. 74, 75

344. *On Certain Developable Surfaces* 267
Quart. Math. Journ. t. VI. (1864), pp. 108—126

345. *On the Inflexions of the Cubical Divergent Parabolas* . . 284
Quart. Math. Journ. t. VI. (1864), pp. 199—203

346. *Note on an expression for the Resultant of two Binary Cubics* 289
Quart. Math. Journ. t. VI. (1864), pp. 380—382

347. *On the Notion and Boundaries of Algebra* 292
Quart. Math. Journ. t. VI. (1864), pp. 382—384

348. *On the Theory of Involution* 295
Camb. Phil. Trans. t. XI. Part I. (1866), pp. 21—38

349. *On a case of the Involution of Cubic Curves.* . . . 313
Camb. Phil. Trans. t. XI. Part I. (1866), pp. 39—80

350. *On the Classification of Cubic Curves* 354
Camb. Phil. Trans. t. XI. Part I. (1866), pp. 81—128

351. *On Cubic Cones and Curves* 401
Camb. Phil. Trans. t. XI. Part I. (1866), pp. 129—144

352. *Suite des recherches sur l'élimination et la théorie des courbes* 416
Crelle, t. LXIV. (1865), pp. 167—171

353. *Note sur la surface du quatrième ordre de Steiner* . . 421
Crelle, t. LXIV. (1865), pp. 172—174

354. *Note sur les singularités supérieures des courbes planes.* . 424
Crelle, t. LXIV. (1865), pp. 369—371

355. *Sur un théorème relatif à huit points situés sur une conique.* 427
Crelle, t. LXV. (1866), pp. 180—184

356. *Sur un cas particulier de la surface du quatrième ordre avec seize points singuliers* 431
Crelle, t. LXV. (1866), pp. 284—291

357. *A Supplementary Memoir on the Theory of Matrices* . . 438
Phil. Trans. t. CLVI. (for 1866), pp. 25—35

PAGE

358. *Addition to the Memoir on Tschirnhausen's Transformation* . 449
Phil. Trans. t. CLVI. (for 1866), pp. 97—100

359. *A Supplementary Memoir on Caustics* 454
Phil. Trans. t. CLVII. (for 1867), pp. 7—16

360. *Note on a Quartic Surface* 465
Phil. Mag. t. XXIX. (1865), pp. 19—22

361. *On Quartic Curves* 468
Phil. Mag. t. XXIX. (1865), pp. 105—108

362. *Note on Lobatschewsky's Imaginary Geometry* 471
Phil. Mag. t. XXIX. (1865), pp. 231—233

363. *On the theory of the Evolute* 473
Phil. Mag. t. XXIX. (1865), pp. 344—350

364. *On a Theorem relating to Five Points in a plane* . . . 480
Phil. Mag. t. XXIX. (1865), pp. 460—464

365. *On the Intersections of a Pencil of four lines by a Pencil of two lines* 484
Phil. Mag. t. XXIX. (1865), pp. 501—503

366. *Note on the Projection of the Ellipsoid* 487
Phil. Mag. t. XXX. (1865), pp. 50—52

367. *On a Triangle in-and-circumscribed to a Quartic Curve* . 489
Phil. Mag. t. XXX. (1865), pp. 340—342

368. *On a problem of Geometrical Permutation* 493
Phil. Mag. t. XXX. (1865), pp. 370—372

369. *On a property of Commutants* 495
Phil. Mag. t. XXX. (1865), pp. 411—413

370. *On the signification of an elementary formula of Solid Geometry* 498
Phil. Mag. t. XXX. (1865), pp. 413, 414

371. *On a Formula for the intersections of a Line and Conic, and on an Integral Formula connected therewith* . . 500
Quart. Math. Journ. t. VII. (1866), pp. 1—6

372. *On the Reciprocation of a Quartic Developable* . . . 505
Quart. Math. Journ. t. VII. (1866), pp. 87—92

373. *On a Special Sextic Developable* 511
Quart. Math. Journ. t. VII. (1866), pp. 105—113

374. *On the Higher Singularities of a Plane Curve* . . . 520
Quart. Math. Journ. t. VII. (1866), pp. 212—223

30   Contents of Volume V.

PAGE

375. *Notes on Polyhedra* 529
Quart. Math. Journ. t. vii. (1866), pp. 304—316

376. *Théorème relatif à l'équilibre de quatre forces*   . . . 540
Comptes Rendus, t. lxi. (1865), pp. 829, 830

377. *Note sur la correspondance de deux points sur une courbe* . 542
Comptes Rendus, t. lxii. (1866), pp. 586—590

378. *Report of a Committee appointed by the British Association to consider the formation of a Catalogue of Philosophical Memoirs (A. Cayley, R. Grant, G. G. Stokes)* . . . 546
Report of the British Association (1856), pp. 463, 464

379. *Notices of Communications to the British Association for the Advancement of Science*   549
Brit. Assoc. Reports, Notices and Abstracts of Communications to the Sections (1854 to 1864)

380. *Note on the Rectangular Hyperbola*   554
Oxford, Camb. and Dubl. Messenger of Mathematics, t. i. (1862), p. 77

381. *Note on Bezout's Method of Elimination* 555
Oxford, Camb. and Dubl. Messenger of Mathematics, t. ii. (1864), pp. 88, 89

382. *Note on the Tetrahedron* 557
Oxford, Camb. and Dubl. Messenger of Mathematics, t. iii. (1866), pp. 8—10

383. *Problems and Solutions* 560
Mathematical Questions with their Solutions from the Educational Times, vols. i. to iv. (1863 to 1865); for contents, see p. 612

Notes and References to papers in Volume V. 613

VOLUME VI.

PAGE

384. *On the Transformation of Plane Curves* 1
Proc. London Math. Society, t. I. (1865—66), No. III. pp. 1—11

385. *On the Correspondence of Two Points on a Curve* . . . 9
Proc. London Math. Society, t. I. (1865—66), No. VII. pp. 1—7

386. *On the Logarithms of Imaginary Quantities* 14
Proc. London Math. Society, t. II. (1866—69), pp. 50—54

387. *Notices of Communications to the London Mathematical Society* 19
Proc. London Math. Society, t. II. (1866—69), pp. 6—7, 25—26,
29, 61—63, 103—104, 123—125

388. *Note on the Composition of Infinitesimal Rotations* . . . 24
Quart. Math. Journ. t. VIII. (1867), pp. 7—10

389. *On a Locus derived from Two Conics* 27
Quart. Math. Journ. t. VIII. (1867), pp. 77—84

390. *Theorem relating to the four Conics which touch the same
two lines and pass through the same four points* . . . 35
Quart. Math. Journ. t. VIII. (1867), pp. 162—167

391. *Solution of a Problem of Elimination* 40
Quart. Math. Journ. t. VIII. (1867), pp. 183—185

392. *On the Conics which pass through two given Points and touch
two given Lines* 43
Quart. Math. Journ. t. VIII. (1867), pp. 211—219

393. *On the Conics which touch three given Lines and pass through
a given Point.* 51
Quart. Math. Journ. t. VIII. (1867), pp. 220—222

394. *On a Locus in relation to the Triangle* 53
Quart. Math. Journ. t. VIII. (1867), pp. 264—277

395. *Investigations in connexion with Casey's Equation* . . . 65
Quart. Math. Journ. t. VIII. (1867), pp. 334—341

PAGE

396. *On a certain Envelope depending on a Triangle inscribed in a Circle*   72
Quart. Math. Journ. t. IX. (1868), pp. 31—41 and 175—176

397. *Specimen Table* $M \equiv a^{\alpha}b^{\beta}$ (*Mod. N*) *for any prime or composite Modulus*   83
Quart. Math. Journ. t. IX. (1868), pp. 95, 96 and plate

398. *On a certain Sextic Developable and Sextic Surface connected therewith*   87
Quart. Math. Journ. t. IX. (1868), pp. 129—142 and 373—376

399. *On the Cubical Divergent Parabolas*   101
Quart. Math. Journ. t. IX. (1868), pp. 185—189

400. *On the Cubic Curves inscribed in a given Pencil of Six Lines*   105
Quart. Math. Journ. t. IX. (1868), pp. 210—221

401. *A Notation of the Points and Lines in Pascal's Theorem* .   116
Quart. Math. Journ. t. IX. (1868), pp. 268—274

402. *On a Singularity of Surfaces*   123
Quart. Math. Journ. t. IX. (1868), pp. 332—338

403. *On Pascal's Theorem*   129
Quart. Math. Journ. t. IX. (1868), pp. 348—353

404. *Reproduction of Euler's Memoir of* 1758 *on the Rotation of a Solid Body*   135
Quart. Math. Journ. t. IX. (1868), pp. 361—373

405. *An Eighth Memoir on Quantics*   147
Phil. Trans. t. CLVII. (for 1867), pp. 513—554

406. *On the Curves which satisfy given Conditions*   191
Phil. Trans. t. CLVIII. (for 1868), pp. 75—143

407. *Second Memoir on the Curves which satisfy given Conditions; the Principle of Correspondence*   263
Phil. Trans. t. CLVIII. (for 1868), pp. 145—172

408. *Addition to Memoir on the Resultant of a System of two Equations*   292
Phil. Trans. t. CLVIII. (for 1868), pp. 173—180

409. *On the Conditions for the existence of three equal Roots or of two pairs of equal Roots of a Binary Quartic or Quintic*   300
Phil. Trans. t. CLVIII. (for 1868), pp. 577—588

PAGE

410. *A Third Memoir on Skew Surfaces, otherwise Scrolls* . . 312
Phil. Trans. t. CLIX. (for 1869), pp. 111—126

411. *A Memoir on the Theory of Reciprocal Surfaces* . . . 329
Phil. Trans. t. CLIX. (for 1869), pp. 201—229

412. *A Memoir on Cubic Surfaces* 359
Phil. Trans. t. CLIX. (for 1869), pp. 231—326

413. *A Memoir on Abstract Geometry* 456
Phil. Trans. t. CLX. (for 1870), pp. 51—63

414. *On Polyzomal Curves, otherwise the Curves* $\sqrt{U}+\sqrt{V}+$ &c. $=0$. 470
Trans. R. Soc. Edinburgh, t. XXV. (for 1868), pp. 1—110

415. *Corrections and Additions to the Memoir on the Theory of Reciprocal Surfaces* 577
Phil. Trans. t. CLXII. (for 1872), pp. 83—87

416. *On the Theory of Reciprocal Surfaces* 582
Addition to Salmon's Analytic Geometry of Three Dimensions, 4th ed. (1882), pp. 592—604

Notes and References to papers in Volume VI. 593
Portrait . . . . *to face Title.*

34

VOLUME VII.

PAGE

417. *On the Locus of the Foci of the Conics which pass through Four Given Points* 1
Phil. Mag. t. XXXII. (1866), pp. 362—365

418. *A Remark on Differential Equations* 5
Phil. Mag. t. XXXII. (1866), pp. 379—381

419. *A Theorem on Differential Operators* 8
Phil. Mag. t. XXXII. (1866), pp. 461—472

420. *On Riccati's Equation* 9
Phil. Mag. t. XXXVI. (1868), pp. 348—351

421. *Note on the Solvibility of Equations by means of Radicals* . 13
Phil. Mag. t. XXXVI. (1868), pp. 386, 387

422. *On the Geodesic Lines on an Oblate Spheroid* 15
Phil. Mag. t. XL. (1870), pp. 329—340

423. *On the Plane Representation of a Solid Figure* . . . 26
Phil. Mag. t. XLI. (1871), pp. 286—290

424. *On the Attraction of a Terminated Straight Line* . . . 31
Phil. Mag. t. XLI. (1871), pp. 358—360

425. *Note on the Geodesic Lines on an Ellipsoid* 34
Phil. Mag. t. XLI. (1871), pp. 534, 535

*426. *On a supposed New Integration of Differential Equations of the Second Order* 36
Phil. Mag. t. XLII. (1871), pp. 197—199

427. *On Gauss' Pentagramma Mirificum* 37
Phil. Mag. t. XLII. (1871), pp. 311, 312

*428. *Note sur la correspondance de deux points sur une courbe* . 39
Comptes Rendus, t. LXII. (1866), pp. 586—590

*429. *Sur les Coniques déterminées par cinq conditions de contact avec une courbe donnée* 40
Comptes Rendus, t. LXIII. (1866), pp. 9—12

PAGE

430. *Note sur quelques formules de M. E. de Jonquières, relatives aux Courbes qui satisfont à des conditions données* . .   41
Comptes Rendus, t. LXIII. (1866), pp. 666—670

431. *Sur la transformation cubique d'une fonction elliptique* . .   44
Comptes Rendus, t. LXIV. (1867), pp. 560—563

432. *Théorème relatif à la théorie des substitutions*   47
Comptes Rendus, t. LXVII. (1868), pp. 784, 785

433. *Sur les surfaces tétraédrales*   48
Notes to De la Gournerie, *Recherches sur les Surfaces réglées tétraé-drales symétriques*, 8vo. Paris, 1867

434. *On Certain Skew Surfaces, otherwise Scrolls*   54
Camb. Phil. Trans. t. XI. Part II. (1869), pp. 277—289

435. *On the Six Coordinates of a Line*   66
Camb. Phil. Trans. t. XI. Part II. (1869), pp. 290—323

436. *On a Certain Sextic Torse*   99
Camb. Phil. Trans. t. XI. Part III. (1871), pp. 507—523

*437. *Démonstration nouvelle du théorème de M. Casey par rapport aux cercles qui touchent à trois cercles donnés* . . .   115
Annali di Matematica, t. I. (1867), pp. 132—134

438. *Note sur quelques torses sextiques*   116
Annali di Matematica, t. II. (1868), pp. 99, 100

439. *Addition à la Note sur quelques torses sextiques* . . .   118
Annali di Matematica, t. II. (1868), pp. 219—221

440. *Note sur une transformation géométrique*   121
Journ. der Mathem. (Crelle), t. LXVII. (1867), pp. 95, 96

441. *Note sur l'algorithme des tangentes doubles d'une courbe du quatrième ordre*   123
Journ. der Mathem. (Crelle), t. LXVIII. (1868), pp. 176—179

442. *Note sur la surface du quatrième ordre douée de seize points singuliers et de seize plans singuliers*   126
Journ. der Mathem. (Crelle), t. LXXIII. (1871), pp. 292, 293

443. *Note on the solution of the Quartic Equation $aU + 6\beta H = 0$* .   128
Math. Ann. t. I. (1869), pp. 54, 55

444. *On the Centro-surface of an Ellipsoid*   130
Proc. Lond. Math. Society, t. III. (1869—1871), pp. 16—18

445. *A Memoir on Quartic Surfaces*   133
Proc. Lond. Math. Society, t. III. (1869—1871), pp. 19—69

PAGE

446. *On the Mechanical Description of a Nodal Bicircular Quartic*   182
Proc. Lond. Math. Society, t. III. (1869—1871), pp. 100—106

447. *On the Rational Transformation between Two Spaces* .   .   189
Proc. Lond. Math. Society, t. III. (1869—1871), pp. 127—180

448. *Note on the Cartesian with Two Imaginary Axial Foci* .   .   241
Proc. Lond. Math. Society, t. III. (1869—1871), pp. 181, 182

449. *Sketch of recent researches upon Quartic and Quintic Sur-*
faces   244
Proc. Lond. Math. Society, t. III. (1869—1871), pp. 186—195

450. *Note on the Theory of the Rational Transformation between*
Two Planes, and on Special Systems of Points   .   .   253
Proc. Lond. Math. Society, t. III. (1869—1871), pp. 196—198

451. *A Second Memoir on Quartic Surfaces*   256
Proc. Lond. Math. Society, t. III. (1869—1871), pp. 198—202

452. *On an Analytical Theorem from a new point of view* .   .   261
Proc. Lond. Math. Society, t. III. (1869—1871), pp. 220, 221

453. *On a Problem in the Calculus of Variations*   263
Proc. Lond. Math. Society, t. III. (1869—1871), pp. 221, 222

454. *A Third Memoir on Quartic Surfaces*   264
Proc. Lond. Math. Society, t. III. (1869—1871), pp. 234—266

455. *On Plücker's Models of certain Quartic Surfaces*   .   .   .   298
Proc. Lond. Math. Society, t. III. (1869—1871), pp. 281—285

456. *Note on the Discriminant of a Binary Quartic*   .   .   .   303
Quart. Math. Journ. t. x. (1870), p. 23

457. *On the Quartic Surfaces* $(*\!\!\!\;\chi U,\ V,\ W)^2 = 0$ 304
Quart. Math. Journ. t. x. (1870), pp. 24—34

458. *On the Anharmonic-Ratio Sextic*   314
Quart. Math. Journ. t. x. (1870), pp. 56, 57

459. *On the Double-Sixers of a Cubic Surface*   316
Quart. Math. Journ. t. x. (1870), pp. 58—71

460. *Note on Mr Frost's paper On the direction of the Lines of*
Curvature in the neighbourhood of an Umbilicus   .   .   330
Quart. Math. Journ. t. x. (1870), pp. 111—113

461. *On the Geometrical Interpretation of the Covariants of a Binary*
Cubic   332
Quart. Math. Journ. t. x. (1870), pp. 148, 149

PAGE

462. *A Ninth Memoir on Quantics* 334
Phil. Trans. t. CLXI. (for 1871), pp. 17—50

463. *Note on a Differential Equation* 354
Mem. Manchester Society, t. II. (1865), pp. 111—114

464. *Note on Plana's Lunar Theory* 357
Monthly Notices R. Ast. Society, t. XXIII. (1862—1863), pp. 211—215

465. *Note on the Lunar Theory* 361
Monthly Notices R. Ast. Society, t. XXV. (1864—1865), pp. 182—189

466. *Second Note on the Lunar Theory* 367
Monthly Notices R. Ast. Society, t. XXV. (1864—1865), pp. 203—207

467. *Expressions for Plana's e, γ in terms of the Elliptic e, γ* . 371
Monthly Notices R. Ast. Society, t. XXV. (1864—1865), pp. 265—271

468. *Addition to Second Note on the Lunar Theory* . . . 375
Monthly Notices R. Ast. Society, t. XXVII. (1866—1867), pp. 267—269

469. *On an Expression for the Angular Distance of two Planets* . 377
Monthly Notices R. Ast. Society, t. XXVII. (1866—1867), pp. 312—315

470. *Note on the Attraction of Ellipsoids* 380
Monthly Notices R. Ast. Society, t. XXIX. (1868—1869), pp. 254—257

471. *Note on the Problem of the Determination of a Planet's Orbit from three observations* 384
Monthly Notices R. Ast. Society, t. XXIX. (1868—1869), pp. 257—259

472. *Note on Lambert's Theorem for Elliptic Motion* . . . 387
Monthly Notices R. Ast. Society, t. XXIX. (1868—1869), pp. 318—320

473. *On the Graphical Construction of the Umbral or Penumbral Curve at any instant during a Solar Eclipse* . . . 390
Monthly Notices R. Ast. Society, t. XXX. (1869—1870), pp. 162—164

474. *On the Geometrical Theory of Solar Eclipses* 392
Monthly Notices R. Ast. Society, t. XXX. (1869—1870), pp. 164—168

475. *On a property of the Stereographic Projection* . . . 397
Monthly Notices R. Ast. Society, t. XXX. (1869—1870), pp. 205—207

476. *On the Determination of the Orbit of a Planet from three observations* 400
Mem. R. Ast. Society, t. XXXVIII. (1870), pp. 17—111

477. *On the Graphical Construction of a Solar Eclipse* . . . 479
Mem. R. Ast. Society, t. XXXIX. (1872), pp. 1—17

478. *On the Geodesic Lines on an Ellipsoid* 493
Mem. R. Ast. Society, t. XXXIX. (1872), pp. 31—53

PAGE

479.   *The Second Part of a Memoir On the Development of the
       Disturbing Function in the Lunar and Planetary Theories*   511
       Mem. R. Ast. Society, t. xxxix. (1872), pp. 55—74

480.   *On the Expression of Delaunay's l, g, h in terms of his
       finally adopted Constants*   528
       Monthly Notices R. Ast. Society, t. xxxii. (1871—1872), pp. 8—16

481.   *On the Expression of M. Delaunay's h + g in terms of his
       finally adopted Constants*   534
       Monthly Notices R. Ast. Society, t. xxxii. (1871—1872), p. 74

482.   *Note on a pair of Differential Equations in the Lunar Theory*   535
       Monthly Notices R. Ast. Society, t. xxxii. (1871—1872), pp. 31, 32

483.   *On a pair of Differential Equations in the Lunar Theory*   .   537
       Monthly Notices R. Ast. Society, t. xxxii. (1871—1872), pp. 201—206

484.   *On the variations of the position of the Orbit in the Planetary
       Theory*   541
       Monthly Notices R. Ast. Society, t. xxxii. (1871—1872), pp. 206—211

485.   *Problems and Solutions*   546
       Mathematical Questions with their Solutions from the Educational
       Times, vols. v. to xii. (1866—1869): for contents, see p. 607

Notes and References to papers in Volume VII.   609

Portrait   .   .   .   .   *.to face Title.*

VOLUME VIII.

PAGE

486. *Note on Dr Glaisher's paper on a theorem in definite integration* 1
Quart. Math. Journ. t. x. (1870), pp. 355, 356

487. *On the quartic surfaces $(*\!\!\!\;\mathcal{X}U, V, W)^2 = 0$* 2
Quart. Math. Journ. t. xi. (1871), pp. 15—25

488. *Note on a relation between two circles* 12
Quart. Math. Journ. t. xi. (1871), pp. 82, 83

489. *On the porism of the in-and-circumscribed polygon, and the (2, 2) correspondence of points on a conic* 14
Quart. Math. Journ. t. xi. (1871), pp. 83—91

490. *On a problem of elimination* 22
Quart. Math. Journ. t. xi. (1871), pp. 99—101

491. *On the quartic surfaces $(*\!\!\!\;\mathcal{X}U, V, W)^2 = 0$* 25
Quart. Math. Journ. t. xi. (1871), pp. 111—113

492. *Note on a system of algebraical equations* 29
Quart. Math. Journ. t. xi. (1871), pp. 132, 133

493. *On evolutes and parallel curves* 31
Quart. Math. Journ. t. xi. (1871), pp. 183—200

494. *Example of a special discriminant* 46
Quart. Math. Journ. t. xi. (1871), pp. 211—213

495. *On the envelope of a certain quadric surface* 48
Quart. Math. Journ. t. xi. (1871), pp. 244—246

496. *Tables of the binary cubic forms for the negative determinants $\equiv 0 \,(\mathrm{mod.}\ 4)$ from -4 to -400; and $\equiv 1 \,(\mathrm{mod.}\ 4)$ from -3 to -99; and for five irregular negative determinants* . 51
Quart. Math. Journ. t. xi. (1871), pp. 246—261

497. *Note on the calculus of logic* 65
Quart. Math. Journ. t. xi. (1871), pp. 282, 283

40 CONTENTS OF VOLUME VIII.

PAGE

498. *On the inversion of a quadric surface* 67
   Quart. Math. Journ. t. xi. (1871), pp. 283—288

499. *On the theory of the curve and torse* 72
   Quart. Math. Journ. t. xi. (1871), pp. 294—317

500. *On a theorem relating to eight points on a conic* . . . 92
   Quart. Math. Journ. t. xi. (1871), pp. 344—346

501. *Review. Pineto's tables of logarithms* 95
   Quart. Math. Journ. t. xi. (1871), pp. 375, 376

502. *On the surfaces divisible into squares by their curves of curvature* 97
   Proc. Lond. Math. Society, t. iv. (1871—1873), pp. 8, 9

503. *On the surfaces each the locus of the vertex of a cone which passes through m given points and touches 6 − m given lines* 99
   Proc. Lond. Math. Society, t. iv. (1871—1873), pp. 11—47

504. *On the mechanical description of certain sextic curves* . . 138
   Proc. Lond. Math. Society, t. iv. (1871—1873), pp. 105—111

505. *On the surfaces divisible into squares by their curves of curvature* 145
   Proc. Lond. Math. Society, t. iv. (1871—1873), pp. 120, 121

506. *On the mechanical description of a cubic curve* . . . 147
   Proc. Lond. Math. Society, t. iv. (1871—1873), pp. 175—178

507. *On the mechanical description of certain quartic curves by a modified oval chuck* 151
   Proc. Lond. Math. Society, t. iv. (1871—1873), pp. 186—190

508. *On geodesic lines, in particular those of a quadric surface* . 156
   Proc. Lond. Math. Society, t. iv. (1871—1873), pp. 191—211

509. *Plan of a curve-tracing apparatus* 179
   Proc. Lond. Math. Society, t. iv. (1871—1873), pp. 345—347

510. *On bicursal curves* 181
   Proc. Lond. Math. Society, t. iv. (1871—1873), pp. 347—352

511. *Addition to the memoir on geodesic lines, in particular those of a quadric surface* 188
   Proc. Lond. Math. Society, t. iv. (1871—1873), pp. 368—380

512. *On a correspondence of points in relation to two tetrahedra* . 200
   Proc. Lond. Math. Society, t. iv. (1871—1873), pp. 396—404

513. *On a bicyclic chuck* 209
   Phil. Mag. t. xliii. (1872), pp. 365—367

PAGE

514. *On the problem of the in-and-circumscribed triangle* . . 212
      Phil. Trans. t. CLXI. (for 1871), pp. 369—412

515. *Sur les courbes aplaties* 258
      Comptes Rendus, t. LXXIV. (1872), pp. 708—712

516. *Sur une surface quartique aplatie* 262
      Comptes Rendus, t. LXXIV. (1872), pp. 1393—1395

517. *Sur les surfaces divisibles en carrés par leurs courbes de
     courbure et sur la théorie de Dupin* 264
      Comptes Rendus, t. LXXIV. (1872), pp. 1445—1449

518. *Sur la condition pour qu'une famille de surfaces données
     puisse faire partie d'un système orthogonal* . . . 269
      Comptes Rendus, t. LXXV. (1872), pp. 177—185, 246—250, 324—330,
      381—385, 1800—1803

519. *On curvature and orthogonal surfaces* 292
      Phil. Trans. t. CLXIII. (for 1873), pp. 229—251

520. *On the centro-surface of an ellipsoid* 316
      Camb. Phil. Trans. t. XII. Part I. (1873), pp. 319—365

521. *On Dr Wiener's model of a cubic surface with 27 real lines;
     and on the construction of a double-sixer* . . . 366
      Camb. Phil. Trans. t. XII. Part I. (1873), pp. 366—383

522. *Note on the theory of invariants* 385
      Math. Ann. t. III. (1871), pp. 268—271

523. *On the transformation of unicursal surfaces* . . . 388
      Math. Ann. t. III. (1871), pp. 469—474

524. *On the deficiency of certain surfaces* 394
      Math. Ann. t. III. (1871), pp. 526—529

525. *An example of the higher transformation of a binary form* . 398
      Math. Ann. t. IV. (1871), pp. 359—361

526. *On a surface of the eighth order* 401
      Math. Ann. t. IV. (1871), pp. 558—560

527. *On a theorem in covariants* 404
      Math. Ann. t. V. (1872), pp. 625—629

528. *On the non-Euclidian geometry* 409
      Math. Ann. t. V. (1872), pp. 630—634

529. *A "Smith's Prize" paper* [1868]; *solutions by Prof. Cayley* . 414
      Oxford, Camb. and Dubl. Messenger of Mathematics, t. IV. (1868),
      pp. 201—226

PAGE

530. *Solution of a Senate-House problem* 436
Oxford, Camb. and Dubl. Messenger of Mathematics, t. v. (1871),
pp. 24—27

531. *A "Smith's Prize" paper* [1869]; *solutions by Prof. Cayley* . 439
Oxford, Camb. and Dubl. Messenger of Mathematics, t. v. (1871),
pp. 41—64

532. *Note on the integration of certain differential equations by
series* 458
Oxford, Camb. and Dubl. Messenger of Mathematics, t. v. (1871),
pp. 77—82

533. *On the binomial theorem, factorials, and derivations* . . 463
Oxford, Camb. and Dubl. Messenger of Mathematics, t. v. (1871),
pp. 102—114

534. *A "Smith's Prize" paper* [1870]; *solutions by Prof. Cayley*. 474
Oxford, Camb. and Dubl. Messenger of Mathematics, t. v. (1871),
pp. 182—203

535. *Note on the problem of envelopes* 491
Messenger of Mathematics, t. I. (1872), pp. 3, 4

536. *Note on Lagrange's demonstration of Taylor's theorem* . . 493
Messenger of Mathematics, t. I. (1872), pp. 22—24

537. *Solutions of a Smith's Prize paper for* 1871 496
Messenger of Mathematics, t. I. (1872), pp. 37—47, 71—77, 89—95

538. *Extract from a letter from Prof. Cayley to Mr C. W. Merrifield* 517
Messenger of Mathematics, t. I. (1872), pp. 87, 88

*539. *Further note on Lagrange's demonstration of Taylor's theorem* 519
Messenger of Mathematics, t. I. (1872), pp. 105, 106

540. *On a property of the torse circumscribed about two quadric
surfaces* 520
Messenger of Mathematics, t. I. (1872), pp. 111, 112

541. *On the reciprocal of a certain equation of a conic* . . . 522
Messenger of Mathematics, t. I. (1872), pp. 120, 121

*542. *Further note on Taylor's theorem* 524
Messenger of Mathematics, t. I. (1872), p. 137

543. *On an identity in spherical trigonometry* 525
Messenger of Mathematics, t. I. (1872), p. 145

544. *On a penultimate quartic curve* 526
Messenger of Mathematics, t. I. (1872), pp. 178—180

PAGE

545. *On the theory of the singular solutions of differential equations of the first order* 529
Messenger of Mathematics, t. II. (1873), pp. 6—12

546. *Theorems in relation to certain sign-symbols* 535
Messenger of Mathematics, t. II. (1873), pp. 17—20

547. *On the representation of a spherical or other surface on a plane: a Smith's Prize dissertation* 538
Messenger of Mathematics, t. II. (1873), pp. 36, 37

548. *On Listing's theorem.* 540
Messenger of Mathematics, t. II. (1873), pp. 81—89

549. *Note on the maxima of certain factorial functions* . . . 548
Messenger of Mathematics, t. II. (1873), pp. 129, 130

550. *Problem and hypothetical theorems in regard to two quadric surfaces* 550
Messenger of Mathematics, t. II. (1873), p. 137

551. *Two Smith's Prize dissertations* [1872] 551
Messenger of Mathematics, t. II. (1873), pp. 145—149

552. *On a differential formula connected with the theory of confocal conics* 556
Messenger of Mathematics, t. II. (1873), pp. 157, 158

553. *Two Smith's Prize dissertations* [1873] 558
Messenger of Mathematics, t. II. (1873), pp. 161—166

554. *An elliptic-transcendent identity* 564
Messenger of Mathematics, t. II. (1873), p. 179

555. *Notices of Communications to the British Association for the Advancement of Science* 565
Brit. Assoc. Reports, Notices and Abstracts of Communications to the Sections (1870, 1871, 1873).

Notes and References to papers in Volume VIII. 569
Prefatory Note vii
Arthur Cayley: biographical notice by the Editor ix
List of courses of lectures delivered by Professor Cayley . . . xlv
Facsimile of the manuscript of his note on p. 569 . . . *Frontispiece*

44

VOLUME IX.

PAGE

556. *On Steiner's surface* 1
       Proc. Lond. Math. Society, t. v. (1873—1874), pp. 14—25

557. *On certain constructions for bicircular quartics.* . . . 13
       Proc. Lond. Math. Society, t. v. (1873—1874), pp. 29—31

558. *A geometrical interpretation of the equations obtained by equating*
       to zero the resultant and the discriminants of two binary
       quantics 16
       Proc. Lond. Math. Society, t. v. (1873—1874), pp. 31—33

559. *[Note on inversion]* 18
       Proc. Lond. Math. Society, t. v. (1873—1874), p. 112

560. *[Addition to Lord Rayleigh's paper "On the numerical calcu-*
       lation of the roots of fluctuating functions"]. . . . 19
       Proc. Lond. Math. Society, t. v. (1873—1874), pp. 123, 124

561. *On the geometrical representation of Cauchy's theorems of root-*
       limitation 21
       Camb. Phil. Trans., t. XII. Part II. (1877), pp. 395—413

562. *On a theorem in maxima and minima: addition [to Mr Walton's*
       paper] by Professor Cayley 40
       Quart. Math. Journ., t. X. (1870), pp. 262, 263

563. *Note on the transformation of two simultaneous equations.* . 42
       Quart. Math. Journ., t. XI. (1871), pp. 266, 267

564. *On a theorem in elimination* 43
       Quart. Math. Journ., t. XII. (1873), pp. 5, 6

565. *Note on the Cartesian* 45
       Quart. Math. Journ., t. XII. (1873), pp. 16—19

566. *On the transformation of the equation of a surface to a set*
       of chief axes 48
       Quart. Math. Journ., t. XII. (1873), pp. 34—38

PAGE

567. *On an identical equation connected with the theory of invariants*   52
Quart. Math. Journ., t. XII. (1873), pp. 115—118

568. *Note on the integrals* $\int_0^\infty \cos x^2\, dx$ *and* $\int_0^\infty \sin x^2\, dx$.   .   .   56
Quart. Math. Journ., t. XII. (1873), pp. 118—126

569. *On the cyclide*   64
Quart. Math. Journ., t. XII. (1873), pp. 148—165

570. *On the superlines of a quadric surface in five-dimensional space*   79
Quart. Math. Journ., t. XII. (1873), pp. 176—180

571. *A demonstration of Dupin's theorem*   84
Quart. Math. Journ., t. XII. (1873), pp. 185—191

572. *Theorem in regard to the Hessian of a quaternary function* .   90
Quart. Math. Journ., t. XII. (1873), pp. 193—197

573. *Note on the* (2, 2) *correspondence of two variables* .   .   .   94
Quart. Math. Journ., t. XII. (1873), pp. 197, 198

574. *On Wronski's theorem*   96
Quart. Math. Journ., t. XII. (1873), pp. 221—228

575. *On a special quartic transformation of an elliptic function*   .   103
Quart. Math. Journ., t. XII. (1873), pp. 266—269

576. *Addition to Mr Walton's paper "On the ray-planes in biaxal crystals"*   107
Quart. Math. Journ., t. XII. (1873), pp. 273—275

577. *Note in illustration of certain general theorems obtained by Dr Lipschitz.*   110
Quart. Math. Journ., t. XII. (1873), pp. 346—349

578. *A memoir on the transformation of elliptic functions*   .   .   113
Phil. Trans., t. CLXIV. (for 1874), pp. 397—456

579. *Address delivered by the President, Professor Cayley, on presenting the Gold Medal of the* [*Royal Astronomical*] *Society to Professor Simon Newcomb*   176
Monthly Notices R. Ast. Society, t. XXXIV. (1873—1874), pp. 224—233

580. *On the number of distinct terms in a symmetrical or partially symmetrical determinant; with an addition* .   .   .   185
Monthly Notices R. Ast. Society, t. XXXIV. (1873—1874), pp. 303—307; p. 335

581. *On a theorem in elliptic motion*   191
Monthly Notices R. Ast. Society, t. XXXV. (1874—1875), pp. 337—339

PAGE

582.	*Note on the Theory of Precession and Nutation* . . .	194
    Monthly Notices R. Ast. Society, t. XXXV. (1874—1875), pp. 340—343

583.	*On spheroidal trigonometry*	197
    Monthly Notices R. Ast. Society, t. XXXVII. (1876—1877), p. 92

584.	*Addition to Prof. R. S. Ball's paper "Note on a transfor-*
    mation of Lagrange's equations of motion in generalised
    coordinates, which is convenient in Physical Astronomy" .	198
    Monthly Notices R. Ast. Society, t. XXXVII. (1876—1877), pp. 269—271

585.	*A new theorem on the equilibrium of four forces acting on a*
    solid body	201
    Phil. Mag., t. XXXI. (1866), pp. 78, 79; Camb. Phil. Soc. Proc., t. I.
    (1866), p. 235

586.	*On the mathematical theory of isomers*	202
    Phil. Mag., t. XLVII. (1874), pp. 444—467

587.	*A Smith's Prize dissertation* [1873].	205
    Messenger of Mathematics, t. III. (1874), pp. 1—4

588.	*Problem* [*on tetrahedra*]	209
    Messenger of Mathematics, t. III. (1874), pp. 50—52

589.	*On residuation in regard to a cubic curve*	211
    Messenger of Mathematics, t. III. (1874), pp. 62—65

590.	*Addition to Prof. Hall's paper "On the motion of a particle*
    toward an attracting centre at which the force is infinite" .	215
    Messenger of Mathematics, t. III. (1874), pp. 149—152

591.	*A Smith's Prize paper and dissertation* [1874]; *solutions and*
    remarks	218
    Messenger of Mathematics, t. III. (1874), pp. 165—183; t. IV.
    (1875), pp. 6—8

592.	*On the Mercator's projection of a skew hyperboloid of revo-*
    lution	237
    Messenger of Mathematics, t. IV. (1875), pp. 17—20

593.	*A Sheepshanks' problem* (1866)	241
    Messenger of Mathematics, t. IV. (1875), pp. 34—36

594.	*On a differential equation in the theory of elliptic functions* .	244
    Messenger of Mathematics, t. IV. (1875), pp. 69, 70

595.	*On a Senate-House problem*	246
    Messenger of Mathematics, t. IV. (1875), pp. 75—78

596.	*Note on a theorem of Jacobi's for the transformation of a double*
    integral	250
    Messenger of Mathematics, t. IV. (1875), pp. 92—94

PAGE

597. *On a differential equation in the theory of elliptic functions* .   253
     Messenger of Mathematics, t. IV. (1875), pp. 110—113

598. *Note on a process of integration*   257
     Messenger of Mathematics, t. IV. (1875), pp. 149, 150

599. *A Smith's Prize dissertation*   259
     Messenger of Mathematics, t. IV. (1875), pp. 157—160

600. *Theorem on the n-th Roots of Unity*   263
     Messenger of Mathematics, t. IV. (1875), p. 171

601. *Note on the Cassinian*   264
     Messenger of Mathematics, t. IV. (1875), pp. 187, 188

602. *On the potentials of polygons and polyhedra*   266
     Proc. Lond. Math. Society, t. VI. (1874—1875), pp. 20—34

603. *On the potential of the ellipse and the circle*   281
     Proc. Lond. Math. Society, t. VI. (1874—1875), pp. 38—58

604. *Determination of the attraction of an ellipsoidal shell on an
     exterior point*   302
     Proc. Lond. Math. Society, t. VI. (1874—1875), pp. 58—67

605. *Note on a point in the theory of attraction*   312
     Proc. Lond. Math. Society, t. VI. (1874—1875), pp. 79—81

606. *On the expression of the coordinates of a point of a quartic
     curve as functions of a parameter*   315
     Proc. Lond. Math. Society, t. VI. (1874—1875), pp. 81—83

607. *A memoir on prepotentials*   318
     Phil. Trans., t. CLXV. (for 1875), pp. 675—774

608. *[Extract from a] Report on Mathematical Tables* . . .   424
     Brit. Assoc. Report, 1873, pp. 3, 4

609. *On the analytical forms called factions*   426
     Brit. Assoc. Report, 1875, Notices of Communications to the Sections,
     p. 10

610. *On the analytical forms called Trees, with application to the
     theory of chemical combinations*   427
     Brit. Assoc. Report, 1875, pp. 257—305

611. *Report on mathematical tables*   461
     Brit. Assoc. Report, 1875, pp. 305—336

612. *Note sur une formule d'intégration indéfinie*   500
     Comptes Rendus, t. LXXVIII. (1874), pp. 1624—1629

613. *On the group of points G_4^1 on a sextic curve with five double
     points*   504
     Math. Ann., t. VIII. (1875), pp 359—362

PAGE

614. *On a problem of projection* 508
      Quart. Math. Journ., t. xiii. (1875), pp. 19—29

615. *On the conic torus* 519
      Quart. Math. Journ., t. xiii. (1875), pp. 127—129

616. *A geometrical illustration of the cubic transformation in elliptic*
    functions 522
      Quart. Math. Journ., t. xiii. (1875), pp. 211—216

617. *On the scalene transformation of a plane curve* . . . 527
      Quart. Math. Journ., t. xiii. (1875), pp. 321—328

618. *On the mechanical description of a Cartesian* 535
      Quart. Math. Journ., t. xiii. (1875), pp. 328—330

619. *On an algebraical operation* 537
      Quart. Math. Journ., t. xiii. (1875), pp. 369—375

620. *Correction of two numerical errors in Sohnke's paper respect-*
    ing modular equations 543
      Crelle, t. lxxxi. (1876), p. 229

621. *On the number of the univalent radicals* $C_n H_{2n+1}$ . . . 544
      Phil. Mag., Ser. 5, t. iii. (1877), pp. 34, 35

622. *On a system of equations connected with Malfatti's problem* . 546
      Proc. Lond. Math. Society, t. vii. (1876), pp. 38—42

623. *On three-bar motion* 551
      Proc. Lond. Math. Society, t. vii. (1876), pp. 136—166

624. *On the bicursal sextic* 581
      Proc. Lond. Math. Society, t. vii. (1876), pp. 166—172

625. *On the condition for the existence of a surface cutting at*
    right angles a given set of lines 587
      Proc. Lond. Math. Society, t. viii. (1877), pp. 53—57

626. *On the general differential equation* $\dfrac{dx}{\sqrt{X}} + \dfrac{dy}{\sqrt{Y}} = 0$, *where* X, Y
    are the same quartic functions of x, y *respectively* . . 592
      Proc. Lond. Math. Society, t. viii. (1877), pp. 184—199

627. *Geometrical illustration of a theorem relating to an irrational*
    function of an imaginary variable 609
      Proc. Lond. Math. Society, t. viii. (1877), pp. 212—214

628. *On the circular relation of Möbius* 612
      Proc. Lond. Math. Society, t. viii. (1877), pp. 220—225

629. *On the linear transformation of the integral* $\displaystyle\int \dfrac{du}{\sqrt{U}}$ . . . 618
      Proc. Lond. Math. Society, t. viii. (1877), pp. 226—229

49

VOLUME X.

PAGE

630. *On an expression for* $1 \pm \sin(2p+1)u$ *in terms of* $\sin u$ . . 1
Messenger of Mathematics, t. v. (1876), pp. 7, 8

631. *Synopsis of the theory of equations* 3
Messenger of Mathematics, t. v. (1876), pp. 39—49

632. *On Aronhold's integration-formula* 12
Messenger of Mathematics, t. v. (1876), pp. 88—90

*633. *Note on Mr Martin's paper "On the integrals of some differentials"* 15
Messenger of Mathematics, t. v. (1876), p. 163

634. *Theorems in trigonometry and on partitions* . . . 16
Messenger of Mathematics, t. v. (1876), p. 164, p. 188

635. *Note on the demonstration of Clairaut's theorem* . . . 17
Messenger of Mathematics, t. v. (1876), pp. 166, 167

636. *On the theory of the singular solutions of differential equations of the first order* 19
Messenger of Mathematics, t. vi. (1877), pp. 23—27

637. *On a differential equation in the theory of elliptic functions* . 24
Messenger of Mathematics, t. vi. (1877), p. 29

638. *On a q-formula leading to an expression for* E_1 . . . 25
Messenger of Mathematics, t. vi. (1877), pp. 63—66

639. *An elementary construction in optics* 28
Messenger of Mathematics, t. vi. (1877), pp. 81, 82

*640. *Further note on Mr Martin's paper* 29
Messenger of Mathematics, t. vi. (1877), pp. 82, 83

641. *On the flexure of a spherical surface* 30
Messenger of Mathematics, t. vi. (1877), pp. 88—90

642. *On a differential relation between the sides of a quadrangle* . 33
Messenger of Mathematics, t. vi. (1877), pp. 99—101

C. XIV. 7

PAGE

643. *On a quartic curve with two odd branches* 36
      Messenger of Mathematics, t. VI. (1877), pp. 107, 108

644. *Note on magic squares* 38
      Messenger of Mathematics, t. VI. (1877), p. 168

645. *A Smith's Prize Paper,* 1877 39
      Messenger of Mathematics, t. VI. (1877), pp. 173—182

646. *On the general equation of differences of the second order* . 47
      Quart. Math. Journ., t. XIV. (1877), pp. 23—25

647. *On the quartic surfaces represented by the equation, symmetrical determinant $= 0$* 50
      Quart. Math. Journ., t. XIV. (1877), pp. 46—52

648. *Algebraical theorem* 57
      Quart. Math. Journ., t. XIV. (1877), p. 53

649. *Addition to Mr Glaisher's Note on Sylvester's paper " Development of an idea of Eisenstein"* 58
      Quart. Math. Journ., t. XIV. (1877), pp. 83, 84

650. *On a quartic surface with twelve nodes* 60
      Quart. Math. Journ., t. XIV. (1877), pp. 103—106

651. *On a special surface of minimum area* 63
      Quart. Math. Journ., t. XIV. (1877), pp. 190—196

652. *On a sextic torse* 68
      Quart. Math. Journ., t. XIV. (1877), pp. 229—235

653. *On a torse depending on the elliptic functions* . . . 73
      Quart. Math. Journ., t. XIV. (1877), pp. 235—241

654. *On certain octic surfaces* 79
      Quart. Math. Journ., t. XIV. (1877), pp. 249—264

655. *A memoir on differential equations* 93
      Quart. Math. Journ., t. XIV. (1877), pp. 292—339

656. *On the theory of partial differential equations* . . . 134
      Mathematische Annalen, t. XI. (1877), pp. 194—198

657. *Note on the theory of elliptic integrals* 139
      Mathematische Annalen, t. XII. (1877), pp. 143—146

658. *On some formulæ in elliptic integrals* 143
      Mathematische Annalen, t. XII. (1877), pp. 369—374

659. *A theorem on groups* 149
      Mathematische Annalen, t. XIII. (1878), pp. 561—565

PAGE

660. *On the correspondence of homographies and rotations* . . 153
Mathematische Annalen, t. XV. (1879), pp. 238—240

661. *On the double ϑ-functions* 155
Proc. Lond. Math. Soc., t. IX. (1878), pp. 29, 30

662. *On the double Θ-functions in connexion with a 16-nodal quartic surface* 157
Crelle's Journal der Mathem., t. LXXXIII. (1877), pp. 210—219

663. *Further investigations on the double ϑ-functions* . . . 166
Crelle's Journal der Mathem., t. LXXXIII. (1877), pp. 220—233

664. *On the 16-nodal quartic surface* 180
Crelle's Journal der Mathem., t. LXXXIV. (1878), pp. 238—241

665. *A memoir on the double ϑ-functions* 184
Crelle's Journal der Mathem., t. LXXXV. (1878), pp. 214—245

666. *Sur un exemple de réduction d'intégrales abéliennes aux fonctions elliptiques* 214
Comptes Rendus, t. LXXXV. (1877), pp. 265—268, 373—376, 426—429, 472—475

667. *On the bicircular quartic—Addition to Professor Casey's memoir: "On a new form of tangential equation"* . . 223
Phil. Trans., t. 167 (for 1877), pp. 441—460

668. *On compound combinations* 243
Proceedings of the Lit. Phil. Soc. Manchester, t. XVI. (1877), pp. 113, 114; Memoirs, ib., Ser. III., t. VI. (1879), pp. 99, 100

669. *On a problem of arrangements* 245
Edin. Roy. Soc. Proc., t. IX. (1878), pp. 338—342

670. *[Note on Mr Muir's solution of a "problem of arrangement"]* 249
Edin. Roy. Soc. Proc., t. IX. (1878), pp. 388—391

671. *On a sibi-reciprocal surface* 252
Berlin, Akad. Monatsber., (1878), pp. 309—313

672. *On the game of mousetrap* 256
Quart. Math. Journ., t. XV. (1878), pp. 8—10

673. *Note on the theory of correspondence* 259
Quart. Math. Journ., t. XV. (1878), pp. 32, 33

674. *Note on the construction of Cartesians* 261
Quart. Math. Journ., t. XV. (1878), p. 34

675. *On the fleflecnodal planes of a surface* 262
Quart. Math. Journ., t. XV. (1878), pp. 49—51

PAGE

676. *Note on a theorem in determinants*   265
     Quart. Math. Journ., t. xv. (1878), pp. 55—57

677. [*Addition to Mr Glaisher's paper "Proof of Stirling's theorem"*]   267
     Quart. Math. Journ., t. xv. (1878), pp. 63, 64

678. *On a system of quadric surfaces*   269
     Quart. Math. Journ., t. xv. (1878), pp. 124, 125

679. *On the regular solids*   270
     Quart. Math. Journ., t. xv. (1878), pp. 127—131

680. *On the Hessian of a quartic surface*   274
     Quart. Math. Journ., t. xv. (1878), pp. 141—144

681. *On the derivatives of three binary quantics*   .   .   .   278
     Quart. Math. Journ., t. xv. (1878), pp. 157—168

682. *Formulæ relating to the right line*   287
     Quart. Math. Journ., t. xv. (1878), pp. 169—171

683. *On the function* arc sin $(x + iy)$ 290
     Quart. Math. Journ., t. xv. (1878), pp. 171—174

684. *On a relation between certain products of differences*   .   .   293
     Quart. Math. Journ., t. xv. (1878), pp. 174, 175

685. *On Mr Cotterill's goniometrical problem*   295
     Quart. Math. Journ., t. xv. (1878), pp. 196—198

686. *On a functional equation*   298
     Quart. Math. Journ., t. xv. (1878), pp. 315—325 ; Proc. Lond. Math.
       Soc., t. ix. (1878), p. 29

687. *Note on the function* $\Im(x) = a^2(c - x) \div \{c(c - x) - b^2\}$.   .   307
     Quart. Math. Journ., t. xv. (1878), pp. 338—340

688. *Geometrical considerations on a solar eclipse* .   .   .   310
     Quart. Math. Journ., t. xv. (1878), pp. 340—347

689. *On the geometrical representation of imaginary variables by
a real correspondence of two planes*   316
     Proc. Lond. Math. Soc., t. ix. (1878), pp. 31—39

690. *On the theory of groups*   324
     Proc. Lond. Math. Soc., t. ix. (1878), pp. 126—133

691. *Note on Mr Monro's paper "On flexure of spaces"*   .   .   331
     Proc. Lond. Math. Soc., t. ix. (1878), pp. 171, 172

692. *Addition to* [578] *memoir on the transformation of elliptic
functions*   333
     Phil. Trans., vol. 169, Part II. (for 1878), pp. 419—424

PAGE

693. *A tenth memoir on quantics* 339
   Phil. Trans., vol. 169, Part II. (for 1878), pp. 603—661

694. *Desiderata and Suggestions* 401

   No. 1. *The theory of groups;*
   American Journal of Mathematics, t. I. (1878), pp. 50—52

   No. 2. *The theory of groups; graphical representation;*
   American Journal of Mathematics, t. I. (1878), pp. 174—176

   No. 3. *The Newton-Fourier imaginary problem;*
   American Journal of Mathematics, t. II. (1879), p. 97

   No. 4. *The mechanical construction of conformable figures;*
   American Journal of Mathematics, t. II. (1879), p. 186

695. *A link-work for x^2: extract from a letter to Mr Sylvester* . 407
   American Journal of Mathematics, t. I. (1878), p. 386

696. *Calculation of the minimum N.G.F. of the binary seventhic* . 408
   American Journal of Mathematics, t. II. (1879), pp. 71—84

697. *On the double ϑ-functions* 422
   Crelle's Journal der Mathem., t. LXXXVII. (1879), pp. 74—81

698. *On a theorem relating to covariants* 430
   Crelle's Journal der Mathem., t. LXXXVII. (1879), pp. 82, 83

699. *On the triple ϑ-functions* 432
   Crelle's Journal der Mathem., t. LXXXVII. (1879), pp. 134—138

700. *On the tetrahedroid as a particular case of the 16-nodal quartic
   surface* 437
   Crelle's Journal der Mathem., t. LXXXVII. (1879), pp. 161—164

701. *Algorithm for the characteristics of the triple ϑ-functions* . 441
   Crelle's Journal der Mathem., t. LXXXVII. (1879), pp. 165—169

702. *On the triple ϑ-functions* 446
   Crelle's Journal der Mathem., t. LXXXVII. (1879), pp. 190—198

703. *On the addition of the double ϑ-functions* 455
   Crelle's Journal der Mathem., t. LXXXVIII. (1880), pp. 74—81

704. *A memoir on the single and double theta-functions* . . . 463
   Phil. Trans., vol. 171, Part III. (for 1880), pp. 897—1002

705. *Problems and Solutions* 566
   Mathematical Questions with their Solutions from the Educational
   Times, vols. XIV. to LXI. (1871—1894); for contents, see p. 615

54

VOLUME XI.

PAGE

706. *On the distribution of electricity on two spherical surfaces* . 1
Phil. Mag., Ser. 5, t. v. (1878), pp. 54—60

707. *On the colouring of maps* 7
Geogr. Soc. Proc., t. I. (1879), pp. 259—261

708. *Note sur la théorie des courbes de l'espace* . . . 9
Assoc. Franç., Compt. Rend., t. IX. (1880), pp. 135—139

709. *On the number of constants in the equation of the surface*
$PS - QR = 0$ 14
Tidsskrift for Mathematik, Ser. 4, t. IV. (1880), pp. 145—148

710. *On a differential equation* 17
Collectanea Mathematica, in memoriam Dominici Chelini, (Milan,
Hoepli, 1881), pp. 17—26

711. *On a diagram connected with the transformation of elliptic
functions* 26
British Association Report, 1881, p. 534

712. *A partial differential equation connected with the simplest case
of Abel's theorem* 27
British Association Report, 1881, pp. 534, 535

713. *Addition to Mr Rowe's "Memoir on Abel's theorem"* . . 29
Phil. Trans., t. CLXXII. (1881), pp. 751—758

714. *Various notes* 37
Messenger of Mathematics, t. VII. (1878), pp. 69 : 115 : 124 : 125

715. *Note on a system of algebraical equations* . . . 39
Messenger of Mathematics, t. VII. (1878), pp. 17, 18

716. *An illustration of the theory of the ϑ-functions* . . . 41
Messenger of Mathematics, t. VII. (1878), pp. 27—32

717. *On the triple theta-functions* 47
Messenger of Mathematics, t. VII. (1878), pp. 48—50

718. *Addition to Mr Genese's paper " On the theory of envelopes "* . 50
Messenger of Mathematics, t. VII. (1878), pp. 62, 63

719. *Suggestion of a mechanical integrator for the calculation of* $\int (Xdx + Ydy)$ *along an arbitrary path* 52
Messenger of Mathematics, t. VII. (1878), pp. 92--95; British Association Report, 1877, pp. 18—20

720. *Note on Arbogast's method of derivations* 55
Messenger of Mathematics, t. VII. (1878), p. 158

721. *Formulæ involving the seventh roots of unity* . . . 56
Messenger of Mathematics, t. VII. (1878), pp. 177—182

722. *A problem in partitions* 61
Messenger of Mathematics, t. VII. (1878), pp. 187, 188

723. *Various notes* 63
Messenger of Mathematics, t. VIII. (1879), pp. 45, 46: 126: 127

724. *On the deformation of the model of a hyperboloid* . . . 66
Messenger of Mathematics, t. VIII. (1879), pp. 51, 52

725. *New formulæ for the integration of* $\dfrac{dx}{\sqrt{X}} + \dfrac{dy}{\sqrt{Y}} = 0$ . . . 68
Messenger of Mathematics, t. VIII. (1879), pp. 60—62

726. *A formula by Gauss for the calculation of* log 2 *and certain other logarithms* 70
Messenger of Mathematics, t. VIII. (1879), pp. 125, 126

727. *Equation of the wave-surface in elliptic coordinates* . . . 71
Messenger of Mathematics, t. VIII. (1879), pp. 190, 191

728. *A theorem in elliptic functions* 73
Proc. Lond. Math. Soc., t. X. (1879), pp. 43—48

729. *On a theorem relating to conformable figures* . . . 78
Proc. Lond. Math. Soc., t. X. (1879), pp. 143—146

730. *[Addition to Mr Spottiswoode's paper " On the twenty-one coordinates of a conic in space"]* 82
Proc. Lond. Math. Soc., t. X. (1879), pp. 194—196

731. *On the binomial equation* $x^p - 1 = 0$; *trisection and quartisection* 84
Proc. Lond. Math. Soc., t. XI. (1880), pp. 4—17

732. *A theorem in spherical trigonometry* 97
Proc. Lond. Math. Soc., t. XI. (1880), pp. 48—50

PAGE

733. *On a formula of elimination*   100
     Proc. Lond. Math. Soc., t. XI. (1880), pp. 139—141

734. *On the kinematics of a plane*   103
     Quart. Math. Journ., t. XVI. (1879), pp. 1—8

735. *Note on the theory of apsidal surfaces*   111
     Quart. Math. Journ., t. XVI. (1879), pp. 109—112

736. *Application of the Newton-Fourier method to an imaginary root of an equation*   114
     Quart. Math. Journ., t. XVI. (1879), pp. 179—185

737. *On a covariant formula*   122
     Quart. Math. Journ., t. XVI. (1879), pp. 224—226

738. *Note on a hypergeometric series*   125
     Quart. Math. Journ., t. XVI. (1879), pp. 268—270

739. *Note on the octahedron function*   128
     Quart. Math. Journ., t. XVI. (1879), pp. 280, 281

740. *On certain algebraical identities*   130
     Quart. Math. Journ., t. XVI. (1879), pp. 281, 282

741. *On a theorem of Abel's relating to a quintic equation* . .   132
     Camb. Phil. Soc. Proc., t. III. (1880), pp. 155—159

742. *On the transformation of coordinates*   136
     Camb. Phil. Soc. Proc., t. III. (1880), pp. 178—184

743. *On the Newton-Fourier problem*   143
     Camb. Phil. Soc. Proc., t. III. (1880), pp. 231, 232

744. *Table of $\Delta^m 0^n \div \Pi\left(m\right)$ up to $m=n=20$*   144
     Camb. Phil. Trans., t. XIII. (1883), pp. 1—4

745. *On the Schwarzian derivative, and the polyhedral functions* .   148
     Camb. Phil. Trans., t. XIII. (1883), pp. 5—68

*746. *Higher Plane Curves*   217
     Salmon's Higher Plane Curves, (3rd ed., 1879), Preface

747. *Note on the degenerate forms of curves*   218
     Salmon's Higher Plane Curves, (3rd ed., 1879), pp. 383—385

748. *On the bitangents of a quartic*   221
     Salmon's Higher Plane Curves, (3rd ed., 1879), pp. 387—389

*749. *Solid Geometry*   224
     Salmon's Treatise on the analytic geometry of three dimensions, (3rd ed., 1874), Preface

PAGE

750. *On the theory of reciprocal surfaces* 225
      Salmon's Treatise on the analytic geometry of three dimensions, (3rd ed., 1874), pp. 539—550

751. *Note on Riemann's paper "Versuch einer allgemeinen Auffass-ung der Integration und Differentiation," Werke, pp. 331—344* 235
      Mathematische Annalen, t. XVI. (1880), pp. 81, 82

752. *On the finite groups of linear transformations of a variable; with a correction* 237
      Mathematische Annalen, t. XVI. (1880), pp. 260—263; 439, 440

753. *On a theorem relating to the multiple theta-functions* . . 242
      Mathematische Annalen, t. XVII. (1880), pp. 115—122

754. *On the connection of certain formulæ in elliptic functions* . 250
      Messenger of Mathematics, t. IX. (1880), pp. 23—25

755. *On the matrix $\begin{pmatrix} a, & b \\ c, & d \end{pmatrix}$, and in connection therewith the function $\dfrac{ax+b}{cx+d}$* 252
      Messenger of Mathematics, t. IX. (1880), pp. 104—109

756. *A geometrical construction relating to imaginary quantities* . 258
      Messenger of Mathematics, t. X. (1881), pp. 1—3

757. *On a Smith's Prize question, relating to potentials* . . . 261
      Messenger of Mathematics, t. XI. (1882), pp. 15—18

758. *Solution of a Senate-House problem.* 265
      Messenger of Mathematics, t. XI. (1882), pp. 23—25

759. *Illustration of a theorem in the theory of equations* . . 268
      Messenger of Mathematics, t. XI. (1882), pp. 111—113

760. *Reduction of $\displaystyle\int \dfrac{dx}{(1-x^3)^{\frac{2}{3}}}$ to elliptic integrals* 270
      Messenger of Mathematics, t. XI. (1882), pp. 142, 143

761. *On the theorem of the finite number of the covariants of a binary quantic* 272
      Quart. Math. Journ., t. XVII. (1881), pp. 137—147

762. *On Schubert's method for the contacts of a line with a surface* 281
      Quart. Math. Journ., t. XVII. (1881), pp. 244—258

763. *On the theorems of the 2, 4, 8, and 16 squares* . . . 294
      Quart. Math. Journ., t. XVII. (1881), pp. 258—276

PAGE

764. *The binomial equation $x^p - 1 = 0$; quinquisection* . . . 314
       Proc. Lond. Math. Soc., t. XII. (1881), pp. 15, 16

765. *On the flexure and equilibrium of a skew surface* . . . 317
       Proc. Lond. Math. Soc., t. XII. (1881), pp. 103—108

766. *On the geodesic curvature of a curve on a surface* . . . 323
       Proc. Lond. Math. Soc., t. XII. (1881), pp. 110—117

767. *On the Gaussian theory of surfaces* 331
       Proc. Lond. Math. Soc., t. XII. (1881), pp. 187—192

768. *Note on Landen's theorem* 337
       Proc. Lond. Math. Soc., t. XIII. (1882), pp. 47, 48

769. *On a formula relating to elliptic integrals of the third kind* . 340
       Proc. Lond. Math. Soc., t. XIII. (1882), pp. 175, 176

770. *On the 34 concomitants of the ternary cubic* 342
       American Journal of Mathematics, t. IV. (1881), pp. 1—15

771. *Specimen of a literal table for binary quantics, otherwise a*
       partition table 357
       American Journal of Mathematics, t. IV. (1881), pp. 248—255

772. *On the analytical forms called trees* 365
       American Journal of Mathematics, t. IV. (1881), pp. 266—268

773. *On the 8-square imaginaries* 368
       American Journal of Mathematics, t. IV. (1881), pp. 293—296

774. *Tables for the binary sextic* 372
       American Journal of Mathematics, t. IV. (1881), pp. 379—384

775. *Tables of covariants of the binary sextic* 377
       Written in 1894 : now first published.

776. *On the Jacobian sextic equation* 389
       Quart. Math. Journ., t. XVIII. (1882), pp. 52—65

777. *A solvable case of the quintic equation* 402
       Quart. Math. Journ., t. XVIII. (1882), pp. 154—157

778. *[Addition to Mr Hudson's paper "On equal roots of equations"]* 405
       Quart. Math. Journ., t. XVIII. (1882), pp. 226—229

779. *[Note on Mr Jeffery's paper "On certain quartic curves,*
       which have a cusp at infinity, whereat the line at infinity
       is a tangent"] 408
       Proc. Lond. Math. Soc., t. XIV. (1883), p. 85

780. *[Addition to Mr Hammond's paper "Note on an exceptional*
       case in which the fundamental postulate of Professor
       Sylvester's theory of tamisage fails"] 409
       Proc. Lond. Math. Soc., t. XIV. (1883), pp. 88—91

PAGE

781. *On the automorphic transformation of the binary cubic function* 411
Proc. Lond. Math. Soc., t. xiv. (1883), pp. 103—108

782. *On Monge's "Mémoire sur la théorie des déblais et des remblais"* 417
Proc. Lond. Math. Soc., t. xiv. (1883), pp. 139—142

783. *On Mr Wilkinson's rectangular transformation* . . . 421
Proc. Lond. Math. Soc., t. xiv. (1883), pp. 222—229

784. *Presidential Address to the British Association, Southport, September* 1883 429
British Association Report, 1883, pp. 3—37

785. *Curve* 460
Encyclopædia Britannica, 9th ed., t. vi. (1878), pp. 716—728

786. *Equation* 490
Encyclopædia Britannica, 9th ed., t. viii. (1878), pp. 497—509

787. *Function* 522
Encyclopædia Britannica, 9th ed., t. ix. (1879), pp. 818—824

788. *Galois* 543
Encyclopædia Britannica, 9th ed., t. x. (1879), p. 48

789. *Gauss* 544
Encyclopædia Britannica, 9th ed., t. x. (1879), p. 116

790. *Geometry (analytical)* 546
Encyclopædia Britannica, 9th ed., t. x. (1879), pp. 408—420

791. *Landen* 583
Encyclopædia Britannica, 9th ed., t. xiv. (1882), p. 271

792. *Locus* 585
Encyclopædia Britannica, 9th ed., t. xiv. (1882), pp. 764, 765

793. *Monge* 586
Encyclopædia Britannica, 9th ed., t. xvi. (1883), pp. 738, 739

794. *Numbers (partition of)* 589
Encyclopædia Britannica, 9th ed., t. xvii. (1884), p. 614

795. *Numbers (theory of)* 592
Encyclopædia Britannica, 9th ed., t. xvii. (1884), pp. 614—624

796. *Series* 617
Encyclopædia Britannica, 9th ed., t. xxi. (1886), pp. 677—682

797. *Surface* 628
Encyclopædia Britannica, 9th ed., t. xxii. (1887), pp. 668—672

798. *Wallis (John)* 640
Encyclopædia Britannica, 9th ed., t. xxiv. (1888), pp. 331, 332

Portrait . . . . *To face Title.*

60

VOLUME XII.

PAGE

799. *On curvilinear coordinates* 1
        Quart. Math. Journ., t. XIX. (1883), pp. 1—22

800. *Note on the standard solutions of a system of linear equations* 19
        Quart. Math. Journ., t. XIX. (1883), pp. 38—40

801. *On seminvariants* 22
        Quart. Math. Journ., t. XIX. (1883), pp. 131—138

802. *Note on Captain MacMahon's paper "On the differential*
    equation $X^{-\frac{1}{3}}dx + Y^{-\frac{1}{3}}dy + Z^{-\frac{1}{3}}dz = 0$ " 30
        Quart. Math. Journ., t. XIX. (1883), pp. 182—184

803. *On Mr Anglin's formula for the successive powers of the root*
    of an algebraical equation 33
        Quart. Math. Journ., t. XIX. (1883), pp. 223, 224

804. *On the elliptic-function solution of the equation $x^3 + y^3 - 1 = 0$* . 35
        Camb. Phil. Soc. Proc., t. IV. (1883), pp. 106—109

805. *Note on Abel's theorem* 38
        Camb. Phil. Soc. Proc., t. IV. (1883), pp. 119—122

806. *Determination of the order of a surface* 42
        Messenger of Mathematics, t. XII. (1883), pp. 29—32

807. *A proof of Wilson's theorem* 45
        Messenger of Mathematics, t. XII. (1883), p. 41

808. *Note on a form of the modular equation in the transform-*
    ation of the third order 46
        Messenger of Mathematics, t. XII. (1883), pp. 173, 174

809. *Schröter's construction of the regular pentagon* 47
        Messenger of Mathematics, t. XII. (1883), p. 177

810. *Note on a system of equations* 48
        Messenger of Mathematics, t. XII. (1883), pp. 191, 192

PAGE

811. *On the linear transformation of the theta-functions* . . 50
Messenger of Mathematics, t. XIII. (1884), pp. 54—60

812. *On Archimedes' theorem for the surface of a cylinder* . . 56
Messenger of Mathematics, t. XIII. (1884), pp. 107, 108

813. [*Note on Mr Griffiths' paper " On a deduction from the elliptic-integral formula $y = \sin (A + B + C + ...)$"*] . . 58
Proc. Lond. Math. Soc., t. XV. (1884), p. 81

814. *On double algebra* 60
Proc. Lond. Math. Soc., t. XV. (1884), pp. 185—197

815. *The binomial equation $x^p - 1 = 0$; quinquisection. Second part* 72
Proc. Lond. Math. Soc., t. XVI. (1885), pp. 61—63

816. *On the bitangents of a plane quartic* 74
Crelle's Journal der Mathem., t. XCIV. (1883), pp. 93—115; Camb. Phil. Soc. Proc., t. IV. (1883), p. 321

817. *On the sixteen-nodal quartic surface* 95
Crelle's Journal der Mathem., t. XCIV. (1883), pp. 270—272

818. *Note on hyperelliptic integrals of the first order* . . . 98
Crelle's Journal der Mathem., t. XCVIII. (1885), pp. 95, 96

819. *On two cases of the quadric transformation between two planes* 100
Johns Hopkins University Circulars, No. 13 (1882), pp. 178, 179

820. *On a problem of analytical geometry* 102
Johns Hopkins University Circulars, No. 15 (1882), p. 209

821. *On the geometrical representation of an equation between two variables* 104
Johns Hopkins University Circulars, No. 15 (1882), p. 210

822. *On associative imaginaries* 105
Johns Hopkins University Circulars, No. 15 (1882), pp. 211, 212

823. *On the geometrical interpretation of certain formulæ in elliptic functions* 107
Johns Hopkins University Circulars, No. 17 (1882), p. 238

824. *Note on the formulæ of trigonometry* 108
Johns Hopkins University Circulars, No. 17 (1882), p. 241

825. *A memoir on the Abelian and Theta Functions* . . . 109
Chapters I to III, American Journal of Mathematics, t. V. (1882), pp. 137—179; Chapters IV to VII, *ib.*, t. VII. (1885), pp. 101—167

PAGE

826. *Note on a partition series* 217
American Journal of Mathematics, t. vi. (1884), pp. 63, 64

827. *On the non-Euclidian plane geometry* 220
Proc. Roy. Soc., t. xxxvii. (1884), pp. 82—102

828. *A memoir on seminvariants* 239
American Journal of Mathematics, t. vii. (1885), pp. 1—25

829. *Tables of the symmetric functions of the roots, to the degree* 10,
for the form $1 + bx + \dfrac{cx^2}{1 \cdot 2} + \ldots = (1 - \alpha x)(1 - \beta x)(1 - \gamma x)\ldots$. 263
American Journal of Mathematics, t. vii. (1885), pp. 47—56

830. *Non-unitary partition tables* 273
American Journal of Mathematics, t. vii. (1885), pp. 57, 58

831. *Seminvariant tables* 275
American Journal of Mathematics, t. vii. (1885), pp. 59—73

832. *Note on an apparent difficulty in the theory of curves, when
the coordinates of a point are given as functions of a
variable parameter* 290
Messenger of Mathematics, t. xiv. (1885), pp. 12—14

833. *On a formula in elliptic functions* 292
Messenger of Mathematics, t. xiv. (1885), pp. 21, 22

834. *On the addition of the elliptic functions* 294
Messenger of Mathematics, t. xiv. (1885), pp. 56—61

835. *On Cardan's solution of a cubic equation* 299
Messenger of Mathematics, t. xiv. (1885), pp. 96, 97

836. *On the quaternion equation* $qQ - Qq' = 0$ 300
Messenger of Mathematics, t. xiv. (1885), pp. 108—112

837. *On the so-called D'Alembert Carnot geometrical paradox* . 305
Messenger of Mathematics, t. xiv. (1885), pp. 113, 114

838. *On the twisted cubics upon a quadric surface.* . . . 307
Messenger of Mathematics, t. xiv. (1885), pp. 129—132

839. *On the matrical equation* $qQ - .Qq' = 0$ 311
Messenger of Mathematics, t. xiv. (1885), pp. 176—178

840. *On Mascheroni's geometry of the compass* 314
Messenger of Mathematics, t. xiv. (1885), pp. 179—181

841. *On a differential operator* 318
Messenger of Mathematics, t. xiv. (1885), pp. 190, 191

842. *On the value of* $\tan(\sin\theta) - \sin(\tan\theta)$ 319
Messenger of Mathematics, t. xiv. (1885), pp. 191, 192

PAGE

843. *On the quadri-quadric curve in connexion with the theory of elliptic functions* 321
Mathematische Annalen, t. xxv. (1885), pp. 152—156

844. *On a theorem relating to seminvariants* 326
Quart. Math. Journ., t. xx. (1885), pp. 212, 213

845. *On the orthomorphosis of the circle into the parabola* . . 328
Quart. Math. Journ., t. xx. (1885), pp. 213—220

846. *A verification in regard to the linear transformation of the theta-functions* 337
Quart. Math. Journ., t. xxi. (1886), pp. 77—84

847. *On the theory of seminvariants.* 344
Quart. Math. Journ., t. xxi. (1886), pp. 92—107

848. *On the transformation of the double theta-functions* . . 358
Quart. Math. Journ., t. xxi. (1886), pp. 142—178

849. *On the invariants of a linear differential equation* . . . 390
Quart. Math. Journ., t. xxi. (1886), pp. 257—261

850. *On linear differential equations* 394
Quart. Math. Journ., t. xxi. (1886), pp. 321—331

851. *On linear differential equations: the theory of decomposition* . 403
Quart. Math. Journ., t. xxi. (1886), pp. 331—335

852. *Note sur le mémoire de M. Picard "Sur les intégrales de différentielles totales algébriques de première espèce"* . . 408
Bull. des Sciences Math., 2me Sér., t. x. (1886), pp. 75—78

853. *Note on a formula for $\Delta^n 0^i / n^i$ when n, i are very large numbers* 412
Proc. Roy. Soc. Edin., t. xiv. (1887), pp. 149—153

854. *An algebraical transformation.* 416
Messenger of Mathematics, t. xv. (1886), pp. 58, 59

855. *Solution of $(a, b, c, d) = (a^2, b^2, c^2, d^2)$* 418
Messenger of Mathematics, t. xv. (1886), pp. 59—61

856. *Note on a cubic equation.* 421
Messenger of Mathematics, t. xv. (1886), pp. 62—64

857. *Analytical geometrical note on the conic* 424
Messenger of Mathematics, t. xv. (1886), p. 192

858. *Comparison of the Weierstrassian and Jacobian elliptic functions* 425
Messenger of Mathematics, t. xvi. (1887), pp. 129—132

PAGE

859. *On the complex of lines which meet a unicursal quartic curve*   428
           Proc. Lond. Math. Soc., t. xvii. (1886), pp. 232—238

860. *On Briot and Bouquet's theory of the differential equation*

$$F\left(u, \frac{du}{dz}\right) = 0$$
           432
           Proc. Lond. Math. Soc., t. xviii. (1887), pp. 314—324

861. *Note on a formula relating to the zero-value of a theta-function*   442
           Crelle's Journal der Mathem., t. c. (1887), pp. 87, 88

862. *Note on the theory of linear differential equations* .   .   .   444
           Crelle's Journal der Mathem., t. c. (1887), pp. 286—295

863. *Note on the theory of linear differential equations* .   .   .   453
           Crelle's Journal der Mathem., t. ci. (1887), pp. 209—213

864. *On Rudio's inverse centro-surface*   457
           Quart. Math. Journ., t. xxii. (1887), pp. 156—158

865. *On multiple algebra*   459
           Quart. Math. Journ., t. xxii. (1887), pp. 270—308

866. *Note on Kiepert's L-equations, in the transformation of elliptic functions*   490
           Mathematische Annalen, t. xxx. (1887), pp. 75—77

867. *Note on the Jacobian sextic equation*   493
           Mathematische Annalen, t. xxx. (1887), pp. 78—84

868. *On the intersection of curves*   500
           Mathematische Annalen, t. xxx. (1887), pp. 85—90

869. *On the transformation of elliptic functions* .   .   .   505
           American Journal of Mathematics, t. ix. (1887), pp. 193—224

870. *On the transformation of elliptic functions (sequel)* .   .   535
           American Journal of Mathematics, t. x. (1888), pp. 71—93

871. *A case of complex multiplication with imaginary modulus arising out of the cubic transformation in elliptic functions*   556
           Proc. Lond. Math. Soc., t. xix. (1888), pp. 300, 301

*872. *On the finite number of the covariants of a binary quantic* .   558
           Mathematische Annalen, t. xxxiv. (1889), pp. 319, 320

873. *System of equations for three circles which cut each other at given angles*   559
           Messenger of Mathematics, t. xvii. (1888), pp. 18—21

874. *Note on the Legendrian coefficients of the second kind* .   .   562
           Messenger of Mathematics, t. xvii. (1888), pp. 21—23

PAGE

875. *On the system of three circles which cut each other at given angles and have their centres in a line* 564
    Messenger of Mathematics, t. xvii. (1888), pp. 60—69

876. *On systems of rays* 571
    Messenger of Mathematics, t. xvii. (1888), pp. 73—78

877. *Note on the two relations connecting the distances of four points on a circle* 576
    Messenger of Mathematics, t. xvii. (1888), pp. 94, 95

878. *Note on the anharmonic ratio equation* 578
    Messenger of Mathematics, t. xvii. (1888), pp. 95, 96

879. *Note on the differential equation* $\dfrac{dx}{\sqrt{(1-x^2)}} + \dfrac{dy}{\sqrt{(1-y^2)}} = 0$ . . 580
    Messenger of Mathematics, t. xviii. (1889), p. 90

880. *Note on the relation between the distance of five points in space* 581
    Messenger of Mathematics, t. xviii. (1889), pp. 100—102

881. *On Hermite's H-product theorem* 584
    Messenger of Mathematics, t. xviii. (1889), pp. 104—107

882. *A correspondence of confocal Cartesians with the right lines of a hyperboloid* 587
    Messenger of Mathematics, t. xviii. (1889), pp. 128—130

883. *Analytical formulæ in regard to an octad of points* . . 590
    Messenger of Mathematics, t. xviii. (1889), pp. 149—152

884. *Note sur les surfaces minima et le théorème de Joachimsthal* . 594
    Comptes Rendus, t. cvi. (1888), pp. 995, 996

885. *On the Diophantine relation,* $y^2 + y'^2 = Square$ 596
    Proc. Lond. Math. Soc., t. xx. (1889), pp. 122—127

886. *On the surfaces with plane or spherical curves of curvature* . 601
    American Journal of Mathematics, t. xi. (1889), pp. 71—98; pp. 293—306

887. *On the theory of groups* 639
    American Journal of Mathematics, t. xi. (1889), pp. 139—157

VOLUME XIII.

PAGE

888. *On a form of quartic surface with twelve nodes* . . . 1
British Association Report, (1886), pp. 540, 541

889. *On a differential equation and the construction of Milner's lamp* 3
Edinburgh Math. Soc. Proc., t. v (1887), pp. 99—101

890. *Note on the hydrodynamical equations* 6
Proc. Roy. Soc. Edin., t. xv (1888), pp. 342—344

891. *On the binodal quartic and the graphical representation of the elliptic functions* 9
Camb. Phil. Soc. Trans., t. xiv (1889), pp. 484—494

892. *Note on the orthomorphic transformation of a circle into itself* 20
Edinburgh Math. Soc. Proc., t. viii (1890), pp. 91, 92

893. *The bitangents of the quintic* 21
Annals of Mathematics, t. v (1890), pp. 109, 110

894. *An investigation by Wallis of his expression for π* . . 22
Quart. Math. Journ., t. xxiii (1889), pp. 165—169

895. *A theorem on trees* 26
Quart. Math. Journ., t. xxiii (1889), pp. 376—378

896. *A transformation in elliptic functions* 29
Quart. Math. Journ., t. xxiv (1890), pp. 259—262

897. *Sur les racines d'une équation algébrique* 33
Comptes Rendus, t. cx (1890), pp. 174—176, 215—218

898. *Sur l'équation modulaire pour la transformation de l'ordre* 11 38
Comptes Rendus, t. cxi (1890), pp. 447—449

899. *Sur les surfaces minima* 41
Comptes Rendus, t. cxi (1890), pp. 953, 954

900. *James Joseph Sylvester* 43
(Scientific Worthies in Nature, xxv); *Nature*, vol. xxxix (1889), pp. 217—219

PAGE

901. *Note on the sums of two series* 49
    Messenger of Mathematics, t. xix (1890), pp. 29—31

902. *On the focals of a quadric surface* 51
    Messenger of Mathematics, t. xix (1890), pp. 113—117

903. *On Latin squares* 55
    Messenger of Mathematics, t. xix (1890), pp. 135—137

904. *Note on reciprocal lines* 58
    Messenger of Mathematics, t. xix (1890), pp. 174, 175

905. *On the equation $x^{17} - 1 = 0$* 60
    Messenger of Mathematics, t. xix (1890), pp. 184—188

906. *Note on Schlaefli's modular equation for the cubic transformation; with a correction* 64
    Messenger of Mathematics, t. xx (1891), pp. 59, 60; 120

907. *Note on the ninth roots of unity* 66
    Messenger of Mathematics, t. xx (1891), p. 63

908. *On two invariants of a quadri-quadric function* . . . 67
    Messenger of Mathematics, t. xx (1891), pp. 68, 69

909. *On a particular case of Kummer's differential equation of the third order* 69
    Messenger of Mathematics, t. xx (1891), pp. 75—79

910. *Note on the involutant of two matrices* 74
    Messenger of Mathematics, t. xx (1891), pp. 136, 137

911. *On an algebraical identity relating to the six coordinates of a line* 76
    Messenger of Mathematics, t. xx (1891), pp. 138—140

912. *On the notion of a plane curve of a given order* . . . 79
    Messenger of Mathematics, t. xx (1891), pp. 148—150

913. *On the epitrochoid* 81
    Messenger of Mathematics, t. xx (1891), pp. 150—158

914. *On a soluble quintic equation* 88
    American Journal of Mathematics, t. xiii (1891), pp. 53—58

915. *On the partitions of a polygon* 93
    Proc. Lond. Math. Soc., t. xxii (1891), pp. 237—262

916. *[Note on a theorem in matrices]* 114
    Proc. Lond. Math. Soc., t. xxii (1891), p. 458

917. *[Note on the theory of rational transformation]* . . . 115
    Proc. Lond. Math. Soc., t. xxii (1891), pp. 475, 476

68 C<small>ONTENTS OF</small> V<small>OLUME</small> XIII.

PAGE

918. *On the substitution-groups for two, three, four, five, six, seven, and eight letters* 117
    Quart. Math. Journ., t. xxv (1891), pp. 71—88, 137—155

919. *On the problem of tactions* 150
    Quart. Math. Journ., t. xxv (1891), pp. 104—127

920. *On orthomorphosis* 170
    Quart. Math. Journ., t. xxv (1891), pp. 203—226

921. *On some problems of orthomorphosis* 191
    Crelle's Journal der Mathem., t. cvii (1891), pp. 262—277

922. *Note on the lunar theory* 206
    Monthly Notices of the Royal Astronomical Society, t. lii (1892),
    pp. 2—5

923. *Note on a hyperdeterminant identity* 210
    Messenger of Mathematics, t. xxi (1892), pp. 131, 132

924. *On the non-existence of a special group of points* . . . 212
    Messenger of Mathematics, t. xxi (1892), pp. 132, 133

925. *On Waring's formula for the sum of the mth powers of the roots of an equation* 213
    Messenger of Mathematics, t. xxi (1892), pp. 133—137

926. *Corrected seminvariant tables for the weights 11 and 12* . 217
    American Journal of Mathematics, t. xiv (1892), pp. 195—200

927. *On Clifford's paper "On syzygetic relations among the powers of linear quantics"* 224
    Proc. Lond. Math. Soc., t. xxiii (1892), pp. 99—104

928. *On the analytical theory of the congruency* 228
    Proc. Lond. Math. Soc., t. xxiii (1892), pp. 185—188

929. *Note on the skew surfaces applicable upon a given skew surface* 231
    Proc. Lond. Math. Soc., t. xxiii (1892), pp. 217—225

930. *Sur la surface des ondes* 238
    Annali di Matematica, Ser. ii, t. xx (1892), pp. 1—18

931. *On some formulæ of Codazzi and Weingarten in relation to the application of surfaces to each other* 253
    Proc. Lond. Math. Soc., t. xxiv (1893), pp. 210—223

932. *On symmetric functions and seminvariants* 265
    American Journal of Mathematics, t. xv (1893), pp. 1—74

PAGE

933. *Tables of pure reciprocants to the weight 8* 333
American Journal of Mathematics, t. xv (1893), pp. 75—77

934. *Note on the so-called quotient G/H in the theory of groups* . 336
American Journal of Mathematics, t. xv (1893), pp. 387, 388

935. *Sur la fonction modulaire $\chi\omega$* 338
Comptes Rendus, t. cxvi (1893), pp. 1339—1343

936. *Note on uniform convergence* 342
Proc. Roy. Soc. Edin., t. xix (1893), pp. 203—207

937. *Note on the orthotomic curve of a system of lines in a plane* 346
Messenger of Mathematics, t. xxii (1893), pp. 45, 46

938. *On two cubic equations* 348
Messenger of Mathematics, t. xxii (1893), pp. 69—71

939. *On a case of the involution $AF + BG + CH = 0$, where A, B, C, F, G, H are ternary quadrics* 350
Messenger of Mathematics, t. xxii (1893), pp. 182—186

940. *On the development of $(1 + n^2 x)^{\frac{m}{n}}$* 354
Messenger of Mathematics, t. xxii (1893), pp. 186—190

941. *Note on the partial differential equation*
$$Rr + Ss + Tt + U(s^2 - rt) - V = 0 \quad . \quad . \quad . \quad . \quad 358$$
Quart. Math. Journ., t. xxvi (1893), pp. 1—5

942. *On seminvariants* 362
Quart. Math. Journ., t. xxvi (1893), pp. 66—69

943. *On reciprocants and differential invariants* 366
Quart. Math. Journ., t. xxvi (1893), pp. 169—194, 289—307

944. *On Pfaff-invariants* 405
Quart. Math. Journ., t. xxvi (1893), pp. 195—205

945. *Note on lacunary functions* 415
Quart. Math. Journ., t. xxvi (1893), pp. 279—281

946. *Note on the theory of orthomorphosis* 418
Quart. Math. Journ., t. xxvi (1893), pp. 282—288

947. *On a system of two tetrads of circles: and other systems of two tetrads* 425
Camb. Phil. Soc. Proc., t. viii (1893), pp. 54—59

PAGE

948. *Report of a Committee appointed for the purpose of carrying on the tables connected with the Pellian equation from the point where the work was left by Degen in 1817* . . 430
British Association Report, (1893), pp. 73—120

949. *On Halphen's characteristic n, in the theory of curves in space* 468
Crelle's Journal der Mathem., t. CXI (1893), pp. 347—352

950. *On the sextic resolvent equations of Jacobi and Kronecker* . 473
Crelle's Journal der Mathem., t. CXIII (1894), pp. 42—49

951. *Non-Euclidian geometry* 480
Camb. Phil. Soc. Trans., t. XV (1894), pp. 37—61

952. *On the kinematics of a plane, and in particular on three-bar motion: and on a curve-tracing mechanism* . . 505
Camb. Phil. Soc. Trans., t. XV (1894), pp. 391—402

953. *On the nine-points circle* 517
Messenger of Mathematics, t. XXIII (1894), pp. 23—25

954. *On the nine-points circle of a plane triangle* 520
Messenger of Mathematics, t. XXIII (1894), pp. 25—27

955. *The numerical value of* $\Pi\left(i\right),\ =\Gamma\left(1+i\right)$ 522
Messenger of Mathematics, t. XXIII (1894), pp. 36—38

956. *On Richelot's integral of the differential equation* $\dfrac{dx}{\sqrt{X}}+\dfrac{dy}{\sqrt{Y}}=0$ 525
Messenger of Mathematics, t. XXIII (1894), pp. 42—47

957. *Illustrations of Sylow's theorems on groups* 530
Messenger of Mathematics, t. XXIII (1894), pp. 59—62

958. *On the surface of the order n which passes through a given cubic curve* 534
Messenger of Mathematics, t. XXIII (1894), pp. 79, 80

959. *Note on Plücker's equations* 536
Messenger of Mathematics, t. XXIV (1895), pp. 23, 24

960. *On the circle of curvature at any point of an ellipse* . . 537
Messenger of Mathematics, t. XXIV (1895), pp. 47, 48

961. *A trigonometrical identity* 538
Messenger of Mathematics, t. XXIV (1895), pp. 49—51

962. *Co-ordinates versus quaternions* 541
Proc. Roy. Soc. Edin., t. XX (1895), pp. 271—275

PAGE

963. *Note on Dr Muir's Paper "A problem of Sylvester's in Elimination"* 545
Proc. Roy. Soc. Edin., t. xx (1895), pp. 306—308

964. *On the nine-points circle of a spherical triangle* . . . 548
Quart. Math. Journ., t. xxvii (1895), pp. 35—39

965. *On the sixty icosahedral substitutions* 552
Quart. Math. Journ., t. xxvii (1895), pp. 236—242

966. *Note on a memoir in Smith's Collected Papers* . . . 558
Bulletin of the American Math. Soc., 2nd Ser., t. i (1895), pp. 94—96

*967. *An elementary treatise on elliptic functions* 560
First edition, 1876, second edition, 1895

INDEX

THIRTEEN VOLUMES.

INDEX.

[*Volumes* are indicated by Roman numerals.
Pages are indicated by Arabic numerals.]

Abbildung: the term, VII, 248; theory, VII, 249—50; (*see also* Transformation of Surfaces).

Abel, N. H.: doubly infinite products, I, 120; inverse elliptic functions, I, 136, 156, 173; modular functions, I, 227; a functional equation, IV, 5—6; elliptic integrals, IV, 63, X, 139; quintic equations, V, 55—61, X, 11, XIII, 88; divergent series, VIII, 494; a theorem of, X, 57; theory of equations, XI, 132—5, 455, 513, 518; elliptic functions, XI, 452; series, XI, 627.

Abelian Functions: Riemann, VI, 2; $p=3$, X, 432—6; early history, XI, 453—4; connected with square roots of sextic and octic functions, XI, 483; the term, XI, 533—4; bitangents of plane quartic, XII, 74; hyperelliptic integrals of first order, XII, 98—9; memoir on theta functions and, XII, 109—216, (introductory, XII, 109; Abel's theorem, XII, 110—20; its proof, XII, 120—31; the major function, XII, 132—48, 149—56; miscellaneous investigations, XII, 157—96; nodal quartic, XII, 196—208; functions T, U, V, Θ, XII, 209—16).

Abelian Integrals: notes on, I, 366—9; and covariants, II, 189—91; Liouville, IV, 546; Riemann, V, 521; reduction to elliptic integrals, X, 214—22; deficiency of ground-curve, XI, 36; first kind of, XII, 38, 408—11; pure theorem for, XII, 110, 112—6, 119—20, 121, 129—30; affected theorem for, XII, 110, 116—7, 121, 130—1, 164—7.

Abel's Theorem: II, 45, 95, XI, 27—8, XII, 30; applied to porisms, IV, 297; Rowe's memoir on, XI, 29—36; note on, XII, 38—41; proof, XII, 120; semi-cubical parabola, XII, 180—6; quadri-quadric curve, XII, 186—9, 292—8; other curves, XII, 189—96.

Absolute: and theory of distance, II, 583—92, 604, V, 550; normals of a conic, IV, 74, 77; theory of evolute, V, 476—9; effect on locus in relation to triangle, VI, 53—64; Cayley's theory of, VIII, xxxvi—vii; evolutes and parallel curves, VIII, 31—44; centro-surface of ellipsoid, VIII, 316, 320; in hypergeometry, VIII, 409—13, XIII, 481—504; the term, XIII, 42; minimal surfaces, XIII, 42.

Abstract Geometry: memoir on, VI, 456—69, VIII, xxxiii, XI, 441—2.

Acceleration: secular (*see* Secular Acceleration).

Acnodal: defined, V, 403, 551, XI, 228.

Acnode: defined, IV, 181, V, 295, 521, VI, 585, XI, 630.

Adams, J. C.: Malfatti's problem, I, 468; moon's mean motion, III, 522, 533—40, 568; lunar theory, VII, 372, VIII, xliii—iv; attraction of ellipsoidal shell, IX, 302; solar eclipses, X, 315.

Addition: of elliptic functions, I, 540—9, 589, XI, 73—7, 454, 530, XII, 294—8; of double theta functions, X, 455—62.

Address: presidential to British Association, XI, 429—59.

Adjoint Curve: IX, 504—7.

Adjoint Linear Form: in quartics, II, 319.

Aggregate: and relation in abstract geometry, VI, 459.

Air: effect on pendulum, IV, 541.

Airy, Sir G. B.: a trigonometrical theorem, IV, 80; roots in algebraic equations, IV, 116, IX, 39.

Algebra: non-commutative, I, 128—31, 301; notion and boundaries, V, 292—4, 620; Cayley founder of modern, VIII, xxx; geometrical illustration of theorems in, IX, 16—7, 21—39; operation connected with covariants, IX, 537—42; expansion theorem, X, 57; system of equations, XI, 39—40; identities, XI, 63—4, 130—1; and time, XI, 443; origin, XI, 445—8; in Greece, XI, 446; and logic, XI, 459, XII, 459; algebraical equations, XI, 506—21; function in, XI, 523—4; Sylvester's work, XIII, 46; Sylvester's principles of universal, XIII, 47; Sylvester on art and, XIII, 48; (*see also* Multiple Algebra).

Algebraic Curves (*see* Curves).

Algebraic Equations (*see* Equations).

Algebraic Theorems: X, 594, 602, 609.

Algebras, Non-commutative: I, 128—31, 301.

Algorithm: for characteristics of triple theta functions, X, 441—5, 452.

Allink: the term, V, 521.

Allotrious: the term, IX, 204.

Alpine Club: Cayley a member, VIII, xi.

Alternant: of operators, XIII, 400—1.

Altitude: of trees, IX, 429—60.

America: Cayley's visit to, VIII, xxi.

American Mathematical Journal: Sylvester's contributions to, XIII, 47.

Ampère, A. M.: inertia, IV, 563, 584; reciprocants, XIII, 366.

Amphigram: the term, VII, 268.

Ampullate: the term, VI, 101.

Anallagmatic: the term, VII, 246.

Analogues: of Pascal's theorem, I, 426, 427.

Analysis: Bernoulli's numbers in, IX, 259—62; (*see also* Combinatory Analysis).

Analytical Geometry (*see* Geometry, analytical).

Analytical Representation of Curves (*see* Representation).

Analytical Theorem: as to Euler's equation, VII, 261—2.

Analytical Theory of conics (*see* Conics).

Anchor Ring: and cyclide, IX, 18.

Angle: interpreted with reference to two points, VI, 497.

Anglin, A. H.: roots of algebraic equations, XII, 33—4.

Angular Distance (*see* Distance).

Anharmonic Ratio: theory of binary quantics, II, 565; sextic equation, VII, 314—5, XII, 578—9.

Anomaly: expansion of the true, III, 139—43; and elliptic motion, IV, 521, 523.

Anti-circle: the term, VIII, 262.

Anti-conic: the term, IX, 65.

Anti-foci: and foci, VII, 567.

Anti-point: the term, VI, 499—500, IX, 65, XIII, 11; problem and solution, VII, 593; orthomorphosis XIII, 184.

Aoust, L.: geodesic curvature, XI, 330.

Aplati (*see* Penultimate Forms).

Apoclastic: the term, XII, 226.

Applicable Surfaces (*see* Skew Surfaces, Surfaces).

Apollonius: tactions, XIII, 152.

Apsidal Surfaces: theory of, XI, 111—3.

Arbelon: the term, XII, 57.

Arbitrary Constants: in mechanical problems, III, 161—5, 200.

Arbogast, L.: method of derivations, II, 257, IV, 265—71, 272—5, 609, XI, 55, 358—60; rule of, VIII, 471, XIII, 267.

Archimedes: and statics, XI, 446; theorem for surface of cylinder, XII, 56—7.

Argand, R.: imaginaries in plane geometry, XII, 460, 468.

Arithmetic: in Greece, XI, 446; Gauss's work, XI, 544.

Arndt, F.: mathematical tables, VIII, 51—5, IX, 490—1.

Aronhold, S. H.: ternary cubic, II, 325, III, 48, IV, 325—30; on lambdaic of binary quartics, II, 550; hyperdeterminants, II, 598—601; incunts, IV, 419; intersection of line and conic, V, 501—4; bitangents of quartic curve, VII, 125; construction of a conic, VII, 593; integration formula, X, 12—4; concomitants of ternary cubic, XI, 342; Abelian function, XII, 157; quadric integral, XII, 162—4, 164—7; cubic transformation, XII, 173—9.

Arrangements: theory of permutations, I, 423—4; triads of seven and fifteen things, I, 481—4, 589, V, 95—7; of numbers, X, 570; latin squares, XIII, 55—7; (*see also* Combinatory Analysis, Permutations, Substitutions).

Art: Cayley's love and practice of, VIII, xxiv; and algebra, XIII, 48.

Askwith, E. H.: substitutions, XIII, 117, 133, 141, 145.

Associative: the term, XII, 461.

Associative Algebra: XII, 303.

Associative Imaginaries: XII, 105—6.

Asteroids: Newcomb on orbits, IX, 176—7.

Astronomy: Cayley's work, VIII, xliii; origin of, XI, 446—7; Gauss's work, XI, 545; transformation of coordinates, XI, 575; Sylvester's work, XIII, 47.

Asymptotes: of algebraic curves, I, 46.

Asymptotic: the term, XIII, 232.

Asyzygetic: covariants and invariants, II, 250; the term, VI, 460—1, VII, 336.

Atomic Theory: Sylvester's work in, XIII, 47.

Attraction and Multiple Integrals: I, 5—12, 13—8, 195—203, 204—6, 285—9, 362—3, 508—10, 586; and theorem of Boole, I, 384—7, 588; and theorem of Jellett, I, 388—91.

Attraction of Ellipsoids: I, 432—4, 511—18, VII, 380—3; Gauss's method, III, 25—8, 149—53; Laplace's, III, 53—65, 567; Rodrigues', III, 149—53.

Attractions: theory, II, 35—9, III, 154—5, XI, 448; of terminated straight line, VII, 31—3; of ellipsoidal shell on exterior point, IX, 302—11; of lens-shaped body, X, 594.

Augmented Equation: XII, 453.

Ausdehnungslehre: Grassmann, XII, 480—9.

Automorphic Transformation: IV, 416, V, 439; of binary cubic function, XI, 411—6.

Autopolar Polyhedra: IV, 185.

Auxesis: XI, 79, 81.

Auxiliars: the term, VI, 156; application to quintics, VI, 186—7.

Auxiliary Equations (*see* Equations).

Axial Systems: of polyhedra, V, 529—39.

Axiom: twelfth of Euclid, XI, 435, XII, 220.

Axis: of inertia, IV, 559—66; cubic surfaces and kinds of, VI, 367; transformation of equation of surface to chief axes, IX, 48—51; radical, XI, 465.

Axonometry: IX, 508—18.

Babbage, C.: homographic function, II, 494; matrices, XI, 252.

Babinet, J.: representation of hemisphere, VIII, 539.

Bacharach, J.: intersection of curves, I, 583, XII, 500.

Baehr, G. F. W.: relative motion, IV, 535, 584.

Ball, Sir R. S.: theory of content, II, 606; roots of quartic, V, 610, VII, 551; dynamical equations, IX, 198—200; non-Euclidian geometry, XIII, 481.

Baltzer, H. R.: book on determinants, IV, 608.

Barriers: the term, X, 320.

Barycentric Calculus: Möbius, XII, 472—3.

Bellavitis, J. L.: equipollences, XII, 473—4.

Beltrami, E.: non-Euclidian plane geometry, XII, 221, 224; orthomorphosis, XIII, 171.

Bernoulli, James: analysis and numbers of, IX, 259—62.

Bertrand, J.: geodesic lines, III, 38; differential equations, III, 164, 189, 203; integrals and mechanical problems, III, 187, 200—3; *Mécanique Analytique*, III, 189—90, 203; wave surface, IV, 432—4; central forces problem, IV, 519—21, 584; problem of two centres, IV, 532, 584; motion of point, IV, 547, 584; problem of three bodies, IV, 547—52, 584, V, 23; curves of curvature on surfaces, VIII, 98; series, XI, 623, 627.

Bessel, F. W.: elliptic motion, IV, 522, 584; precession and nutation, IX, 194—6.

Bezout, E.: determinants, I, 63; elimination, IV, 38—9, V, 159, 555—6; equation of differences, IV, 151, 259.

Bezoutiant: defined, II, 526.

Bezoutic Emanant: defined, II, 525.

Bezoutic Matrix: defined, IV, 607.

Bezoutoid: defined, II, 526.

Biaxial: defined, XIII, 13.

Bibasic: defined, XII, 642.

Bibliography: of covariants and invariants, II, 598—601; of symmetric functions, II, 602—3.

Bicentre: the term, IX, 429—60, X, 599.

Bicircular Quartic, on the: X, 223—42; introductory, X, 223—4; formulæ for fourfold generation, X, 224—6; determination as to reality, X, 226; investigation of elementary *arc* formulæ, X, 226—31; inscribed quadrilateral, X, 231—5.

Bicircular Quartics: and polyzomal curves, VI, 472; foci of, VI, 521—2, 522—3; analytical theory, VI, 528—30; problem and solution, VII, 575, X, 596—8; geometrical construction, IX, 13—5; generation, XIII, 12; transformation of circle, XIII, 185; (*see also* Quartics).

Bicircular Quartic Surfaces: VII, 67, 246; quadric surface inversion, VII, 67.

Bickmore, C. E.: Pellian equation, XIII, 442.

Bicolumn: the term, IX, 28.

Bicorn: the term, VI, 148, 158; equation, VI, 163; form, VI, 164.

Bicursal Curves: VIII, 181—7.

Bicursal Sextic: and binodal quartic, IX, 581.

Bien déterminée: defined, XII, 433.

Bin: the abbreviation for tortuous curves, XIII, 253.

Binary: the term, IV, 604, VI, 466.

Binary Cubics: covariants of, II, 189—91; fifth memoir on quantics, II, 540—5; geometrical representation of covariants of, VII, 332—3; tables of forms, VIII, 51—64; transformation of function, XI, 411—6.

Binary Forms: canonic root, V, 103—5.

Binary Matrices (*see* Matrices).

Binary Quadratic Forms: tables of, V, 141—56, 618.

Binary Quadrics: covariants of, II, 189—91; single, II, 527—9; theories of harmonic relation and of involution, II, 529—40; asyzygetic covariants of, VII, 337—8.

Binary Quantics: canonical form, IV, 43—52, 53; canonical root, V, 103—5; involution, V, 296—301; transformations, VI, 187—90; discriminants of, VII, 303; and covariants, VIII, 566—7, X, 430—1, XI, 272—8; geometrical interpretation, IX, 16—7; factions, IX, 426; derivatives of three, X, 278—86; literal table for, XI, 357—64; finite number of covariants of, XII, 558.

Binary Quartics: covariants of, II, 189—91; fifth memoir on quantics, II, 545—56; theorem of four, VII, 100; generating function of, X, 341.

Binary Quintics: tables of covariants M to W of, II, 282—309; covariants and syzygies of degree 6, VI, 148—53; and sextic, VI, 190; irreducible covariants, VII, 334; tables, VII, 341—6; asyzygetic covariants of, VII, 399—400; $(*\!\!\;\!\,\!(x, \ y)^5$, X, 339—400; theorem relating to covariants, X, 430—1; concomitant system for, XI, 272.

Binary Septic: minimum N.G.F., X, 408—21.

Binary Sextic: and quintic, VI, 190; and squared cubic and cubed quadric, XI, 105; concomitant system for, XI, 272; tables for, XI, 372—6, 377—88.

Binet, J.: determinants, I, 63, 64, 581; variation of arbitrary constants, III, 181; relative motion, IV, 535, 584; inertia, IV, 562—3, 584.

Binodal Quartic: transformed to bicursal sextic, IX, 581; graphical representation of elliptic functions, XIII, 9—19; (*see also* Quartics).

Binode: the term, VI, 330, 335, 361, 362, 583—5.

Binomial Equation: theory of numbers, XI, 84—96.

Binomial Theorem: factorials and derivations, II, 101, 102, VIII, 463—73.

Binormal: the term, XIII, 253.

Biography: of Cayley, VIII, ix—xliv.

Bipartite: the term, VI, 464.

Biplanar-node: the term, VI, 360.

Biplanes: the term, VI, 360—1, 362.

Bipoint-locus: the term, VI, 198.

Biquaternions: Clifford, XI, 458, XIII, 481.

Bisection: theory of numbers, XI, 84—96.

Bitangents: of a plane curve, IV, 186—206; of quartic curve, IV, 342—8, VII, 123—4, X, 244, XI, 221—3, XII, 74—94; of quartic, and triple theta functions, X, 444, 446; of curves, XI, 473—4, 480; of quintic, XIII, 21.

Bitetrad: the term, XIII, 551.

Bitrope: the term, VI, 330, 335, 583—5, 591, XI, 228.

Björling, E. G.: root limitation, IX, 39; difference table, XI, 144.

Blissard, J.: factorials, V, 574.

Blunt: the term in seminvariants, XIII, 265, 287, 291, 304—6.

Bodies, Problem of Three: III, 183; Hamilton's method, III, 97—103; Jacobi's theorem, III, 519—21, V, 23.

Body (*see* Solid Body).

Bolyai, J.: hyperbolic geometry, VIII, xxxvii.

Bondset: the term, XII, 642.

Bonnet, Ossian: problem of two centres, IV, 528, 585; imaginary coordinates, VIII, 569; minimal surfaces, XII, 595; curves of curvature, XII, 601, 629—31; skew surfaces, XIII, 231, 237.

Book-keeping: pamphlet by Cayley, VIII, xxiv.

Boole, G.: linear transformations, I, 80, 92, 93—4, 428, 584; multiple integrals, I, 198, 384—7, 588; attractions, I, 285, 289; theorem of Jellett, I, 388; transformation of elliptic integral, I, 508—10; discriminants, I, 584; theory of permutants, II, 26; theory of attractions, II, 35; probabilities, II, 103, 594—8, V, 80—4, 85; hyperdeterminants, II, 598—601; transformation, III, 129; formulæ for differentiation, IV, 135; differential equations, IV, 515, V, 301; involution, V, 301; invariants, VIII, xxx; calculus of logic, VIII, 65—6; integration of differential equations by series, VIII, 458;

prepotentials, IX, 417—23; difference table, XI, 144; matrices, XI, 256; mathematics and logic, XI, 458—9; series, XI, 627.

Booth, J.: rotation of solid body, IV, 577, 585.

Borchardt, C. W.: symmetric functions, II, 417, 421—3; 16-nodal quartic surface, X, 180; theta functions, X, 464, 499; theorem on trees, XIII, 28.

Bordered Skew Determinants: II, 203.

Boron: trees of, IX, 429—60.

Boundaries: of Algebra, V, 292—4, 620.

Bouquet, J. C.: orthogonal surfaces, VIII, 279, 292; periods of elliptic integrals, IX, 618; conformal representation, XI, 80; imaginary variables, XI, 440; elliptic functions, XI, 534; differential equations of first order, XII, 432—41; functions of complex variables, XIII, 190.

Bour, E.: differential equations, III, 164, 197—8, 204, IV, 515, 585; problem of three bodies, IV, 550—2, 585; Gaussian theory of surfaces, XI, 331—6.

Branch: the term, III, 243, X, 36, XI, 476—7; partial, of a curve, V, 425; of polyzomal curves, VI, 474—6, 481—5; main, of trees, IX, 429—60; theorem on trees, XIII, 26—8.

Branch-points: the term, X, 317.

Brassinne, E.: differential equations, III, 186—7, 203.

Bravais, A.: spherical pendulum, IV, 533, 585; on polyhedra, V, 537, 559.

Bretschneider, C. A.: mathematical tables, IX, 486.

Brianchon, C. J.: theorem of, I, 328, V, 4; rectangular hyperbola, III, 254; reciprocal polars, XI, 465.

Briggs, H.: logarithms, XI, 447.

Brill, A.: transformation of plane curves, VI, 593; transformation, and theory of invariants, VIII, 387; sextic curve, IX, 504—7.

Brinkley, J.: formulæ of, IV, 262.

Brioschi, F.: *Sulla variazione* and *Teorema di Mechanica*, III, 190, 203; *degli integrali di un problema di dinamica*, III, 199—200, 203; peninvariants, IV, 246; text-book on determinants, IV, 608; Schwarzian derivative and polyhedral functions, XI, 149, 151; concomitants of ternary cubic, XI, 342; Jacobian sextic equation, XI, 401; theory of equations, XI, 520; elliptic functions and quintic, XII, 493; transformation of elliptic functions, XII, 505—34, 554—5; wave surfaces, XIII, 238; sextic resolvent equations, XIII, 473, 478, 479.

Briot, A.: rotation of solid body, IV, 574, 585; periods of elliptic integral, IX, 618; conformal representation, XI, 80; imaginary variables, XI, 440; elliptic functions, XI, 534; differential equations of first order, XII, 432—41; functions of complex variables, XIII, 190.

British Association: report on Catalogue of Philosophic Memoirs, V, 546—8, 620; communications to, V, 549—53; Cayley president 1883, VIII, xxi—xxii; communications on in-and-circumscribed triangle, VIII, 565—6; correspondence of points and lines in space, VIII, 566; covariants of binary quantic, VIII, 566—7; families of surfaces, VIII, 567; Mercator's projection of surface of revolution, VIII, 567; report on mathematical tables, IX, 461—99; Presidential address, XI, 429—59.

Bronwin, B.: elliptic functions, I, 118, 119, 127, 132.

Buchheim, A.: parallels of, XIII, 481, 489.

Buée, A. Q.: multiple algebra, XII, 467, 471.

Burg, A.: algebraical theorem, X, 57.

Burnside, W. S.: cubic identity, V, 597.

Cagnoli, A.: spherical trigonometry, IV, 80.

Calculus: of logic, VIII, 65—6; and time, XI, 443—4; of functions, XI, 541—2; of forms; (*see also* Covariants, Invariants, Quantics).

Calculus of Variations: Jacobi on, III, 174; problem in, VII, 263.

Cambridge Mathematical Journal: foundation, VIII, xii.

Campaniform: the term, VI, 101, XI, 479.

Canonical Elements: for disturbed motion, III, 77.

Canonical Forms: of quantics, II, 523, 542, 548; of binary quantics, IV, 43—52, 53; the term, IV, 606—7, XIII, 46; quintics, VI, 153—4, X, 355—62, XII, 27; cubic surfaces, VI, 361; Sylvester's work, XIII, 46; (*see also* Formulæ).

Canonisant: defined, II, 523, IV, 45, 53, 606, 607, V, 103—5.

Capacity: of a curve, I, 583; the term, VII, 140; and postulandum, XIII, 115.

Capitation: and seminvariants, XII, 248—50.

Carbon: trees of, IX, 427—60, 544—5.

Cardan, J.: solution of cubic, V, 54, XII, 299.

Cards: game of mousetrap, III, 8, X, 256—8; (*see also* Permutations).

Carey, F. S.: theory of numbers, XII, 73.

Carnot, L.: distances of points, I, 64, 581; on two tetrahedra in perspective, IX, 209—10; geometrical paradox, XII, 305—6.

Cartesians: cusp, I, 589; oval of Descartes, II, 336, 365—6, 370; equation of, II, 370—4, 376; defined, VI, 524, XI, 481; with two imaginary axial foci, VII, 241—3; and cubic curve, VII, 556; problem and solution, VII, 582; note on, IX, 45—7; construction, IX, 317, 535—6, X, 261; invention of coordinates, XI, 449; explained, XI, 461; scalene transformation, IX, 532—4; correspondence of confocal, with right lines of hyperboloid, XII, 587—9.

Casey, J.: equation of, and tangent circles, VI, 65—71; polyzomal curves, VI, 471—2; bicircular quartics, VI, 540, X, 223—42, XIII, 12; tactions, VI, 543, VII, 115, XIII, 153; circle touching three others, VI, 568—73; cyclides and sphero-quartics, VIII, 262—3; cyclide, IX, 64, 75.

Cassinian: note on, IX, 264—5; the word, X, 608; equation of, X, 608.

Castillon, J. F. S. de: problem and porism, IV, 435—41.

Catalan, E.: attractions, I, 288; integral calculus, I, 313; series, III, 127, XI, 627; curve parallel to ellipse, IV, 152; minimal surfaces, XII, 594—5.

Catalecticant: the term, II, 523, IV, 606, 607.

Catalogue of Scientific Papers: issued by Royal Society, V, 546—8, 620.

Catenary: and epitrochoid, XIII, 87.

Cauchy, A. L.: determinants, I, 63, 64, 66; inverse elliptic functions, I, 148, 157, 174; integral calculus, I, 309, 314; permutations, I, 423; partitions, II, 248, V, 48, VII, 577; differential equations, III, 166, 169; theory of logarithms, III, 208—13, 225; polyhedra, IV, 86—7; singularity of function, IV, 105; curve parallel to ellipse, IV, 152; equations of motion, IV, 514, 585; elliptic motion, IV, 524, 585; inertia, IV, 560, 564, 565, 585; geodesic lines on ellipsoid, VII, 498; geometrical representation of root-limitation, IX, 21—39; series, IX, 61, XI, 627; mathematical tables, IX, 475; theory of equations, X, 6, XI, 504; correspondence, X, 290; monogenous function, XI, 537; theory of numbers, XI, 601; roots of algebraic equations, XIII, 35.

Caustics: by reflection at circles, I, 273—5, II, 118—22, 129; memoirs on, II, 336—80, V, 454—65; tracing of, II, 351—2, 362—3, 368—9, 374.

Cavalieri, B.: quadrature of curves, XI, 641.

Cayley, Arthur: portraits of, frontispiece to vols. VI, VII, XI; manuscript of, frontispiece to vol. VIII; biography of, VIII, IX—xliv; mathematical tables, IX, 479, 487, 491.

Cayleyan: the word, I, 586, II, 381.

Central Forces Problem: IV, 517—21.

Centres: problem of two, III, 104—10, IV, 527, 528—9, 589; cubic, and cones, IV, 173—8, 179—81; of curves defined, VI, 522; four points and conic of, VII, 582; of trees, IX, 429—60; the term, X, 599; of three circles, XI, 552.

Centro-curve: kinds of, VIII, 320.

Centro-surface: of ellipsoid, VIII, 8; the term, VIII, 316; Rudio's inverse, XII, 457—8.

Cesser: points of, IV, 130.

Challis, J.: integration of differential equations, VII, 36.

Chance (*see* Probability).

Characteristic Function: of Hamilton, III, 217; for systems of rays, XII, 571.

Characteristics: logic of, III, 51—2; of Chasles, V, 552; theory, VI, 594, XIII, 468—72; of triple theta functions, X, 441—5.

Chartography: surface representation on plane, VIII, 538—9; colouring of maps, XI, 7—8; map projections, XI, 448.

Chasles, M.: intersections of curves, I, 25—7; Pascal's theorem, I, 45; theorem on correspondence, I, 212; a theorem of, demonstrated, I, 355; analogue of Pascal's theorem, I, 427; transformation of curves, I, 478—80; homography, II, 538; cubic curves, IV, 122, 495; inertia, IV, 561, 586; kinematics of solid body, IV, 580, 586; curves on a quadric, V, 11; on a cubic, V, 19; conics touching curves, V, 31—2, 552; scrolls, V, 169, 201, VI, 328; quartic scrolls, V, 201; cubic curves and cones, V, 401; equilibrium of four forces, V, 540—1; correspondence of points in plane curve, V, 542; contact of conics, V, 552; characteristics, V, 552; on united points, VI, 9; curves which satisfy given conditions, VI, 191, 192, 200—26; principle of correspondence, VI, 264, XI, 482, 485—8; foci of conics, VII, 1; six coordinates of a line, VII, 93; attraction of ellipsoids, VII, 380—3; locus *in plano*, VII, 605; cones satisfying six conditions, VIII, 99; penultimate forms of curves, VIII, 258; theory of duality, XI, 467.

Chemistry: Cayley's interest in, VIII, x; application of trees to, IX, 202—4, 427—60, 544—5.

Chessboard: topology of, X, 609—10.

Chord: angle between normal and bisector, X, 576; of two circles, XI, 552—6.

Christie, J. T.: Cayley's law work, VIII, xiii—iv.

Christoffel, E. B.: orthomorphosis, XIII, 180.

Chrystal, G.: uniform convergence, XIII, 343—4.

Chuck: for quartic curves, VIII, 151—5; for curve-tracing, VIII, 179—80; bicyclic, VIII, 209—11.

Circle: Salmon's equation for orthotomic, III, 48—50; and points, V, 560; and ellipse, V, 561; line and parabola, V, 607; envelope of, V, 610; equation of, VI, 501, XI, 558—61; potential of, IX, 290—301; quadrilateral inscribable in, X, 578; orthomorphosis, XII, 328—36, XIII, 20, 182, 202—5; Wallis's π investigation, XIII, 22—5; transformation into bicircular quartic, XIII, 185; and circumference, the terms, XIII, 194; the nine point, XIII, 517—9, 520—1, 548—51; of curvature of an ellipse, XIII, 537.

Circles: powers of, I, 581; systems of, III, 111—4, X, 566; in-and-circumscribed polygon, IV, 303—8; a pair touching three given, VI, 65—71; involution of four, VI, 505—8; relation between two, VIII, 12—3; equal, VIII, 31; minimum enclosing three points, X, 576; system of 15 connected with icosahedron, XI, 208—12; radical axis, XI, 465; radical centre of three, XI, 552; Mascheroni's geometry of the compass, XII, 314—7; system of three which cut each other at given angles, XII, 559—61, 564—70; the two relations connecting the distances of four points on a circle, XII, 576—7; roots of algebraic equation, XIII, 37; problem of tactions, XIII, 150—69; tetrads of, XIII, 425—9; (*see also* Casey, Orthomorphosis).

Circuit: the word, XI, 480.

Circular: the word, XI, 481.

Circular Cubic: and polyzomal curves, VI, 522—8.

Circular Points: at infinity, VIII, 32.

Circular Relation of Möbius: III, 118—9, IX, 612—7.

Circumference: and circle, the terms, XIII, 194.

Cissoid: the term, XI, 461.

Clairaut, A. C.: lunar theory, IV, 518, 586; demonstration of his theorem, X, 17—8; curves of double curvature, XI, 489.

Classes: of curves, II, 569—83, V, 518, 521, VI, 11, XI, 629.

Classification: of curves, V, 613—7; cubics and quartics, VI, 525; quartic surfaces, VII, 244; nodal quartic surfaces, VII, 264—7; mathematical tables, IX, 424—5; cubic curves, XI, 478—80; quartic curves, XI, 480.

Clausen, T.: Castillon's problem, IV, 435—41.

Clebsch, R. F. A.: Steiner's extension of Malfatti's problem, II, 593; Pfaffian equations, IV, 515; singularities of curves, V, 476—7; developable surfaces, V, 518; Abelian integrals, V, 521; transformation of plane curves, VI, 1—8, 593; Casey's equation, VI, 67; binary quintic and sextic, VI, 190; principle of correspondence, VI, 264; reciprocal surfaces, VI, 356; cubic curve in pencil of six lines, VI, 593—4; centro-surface of ellipsoid, VII, 130, VIII, 316; quartic and quintic surfaces, VII, 245—51, 313; bicursal curves, VIII, 182; transformation of unicursal surfaces, VIII, 389; deficiency of surfaces, VIII, 394; covariants, VIII, 404; expression for elliptic integral of second kind, X, 27; concomitants of ternary cubic, XI, 343; tables for binary sextic, XI, 372; Abelian functions, XII, 102, 109; symbolic notation for quantics, XII, 347; seminvariants, XIII, 362; Pfaff-invariants, XIII, 405—14.

Clifford, W. K.: powers of circles and spheres, I, 581; circle and ellipse, V, 561; conic, V, 562; triangles and polygons, V, 589; rational transformation between two spaces, VII, 222—4; quartic surfaces, VII, 246; compound combinations, X, 243; biquaternions, XI, 458; syzygetic relations among the powers of linear quantics, XIII, 224—7; non-Euclidian geometry, XIII, 481.

Close: partitions of a, V, 62—5, 617; defined, V, 63.

Close-planes: the term, VI, 330, 583—5.

Close-points: the term, VI, 330, 339, 341, 583—5.

Cnicnode: the word, VI, 330, 334, 361, 583—5, XI, 228, 631, 633; quartic and quintic surfaces, VII, 245—51.

Cnictrope: the word, VI, 330, 334, 583—5, 591, XI, 228, 631, 633.

Cobezoutiants: defined, II, 524.

Cobezoutoids: defined, II, 524—5.

Cockle, Sir James: resolvent product, IV, 309—13; on quintics, V, 53; theory of equations, XI, 520; invariants, XII, 390—1; criticoids, XIII, 366, 367—8.

Codazzi, D.: application of surfaces, XIII, 253—64.

Coefficients: of Legendre, I, 375—6; development of $(1+n^2x)^{m/n}$, XIII, 354—7.

Cogredient: the term, IV, 607—8, XIII, 46.

Cohen, A.: relative motion, IV, 537, 586; rotation of solid body, IV, 573, 586.

Cole, F. W.: on simple groups, XIII, 533.

Colink: the term, V, 521.

Collins, M.: Lagrange's theorem, II, 3; systems of circles, X, 566—7.

Colour Groups: diagrams representing, X, 328—330, 403—5; the term, XII, 639—41.

Colours: in maps, XI, 7—8.

Colourset: the term, XII, 641.

Columns: the term, XII, 353.

Comberousse, C.: polyhedra, IV, 609.

Combescure, E.: wave surfaces, XIII, 238.

Combinants: of quantics, II, 322; the term, II, 518, IV, 607—8, XIII, 46; and ternary quadratics, IV, 352.

Combinatory Analysis: magic squares, X, 38; compound combinations, X, 243—4; a problem of arrangements, X, 245—8; note on Mr Muir's solution, X, 249—51; the game of mousetrap, X, 256—8; Sylvester's work in, XIII, 47; (see also **Partitions**).

Commutants: the term, I, 584, II, 19, 26, IV, 594, 599—600, XIII, 46; property of, V, 495—7.

Commutative: the term, XII, 461; (see also **Groups**).

Comoment: non-Euclidian geometry, XIII, 481—9.

Compass: Mascheroni's geometry of the, XII, 314—7.

Complex (*see* Surfaces).

Complex Cone: (cubic) defined, V, 402, 404, 551.

Complexes of Lines: IV, 618; through twisted quartic, XII, 428—31.

Complex Multiplication: in elliptic functions, XII, 556—7; (*see also* Multiplication).

Complex Quantities: logarithms of, VI, 14—8.

Complex Variables: and conformal representation, X, 316—23; Newton-Fourier theorem for, X, 405—6; (*see also* Function, Newton-Fourier theorem).

Composition: of quadratic forms, I, 532; of singularities, V, 619; of rotations, VI, 24—6.

Compound Combinations (*see* Combinatory analysis).

Compound Singularities: V, 525.

Conchoid: the term, XI, 460.

Concomitant: the term, IV, 607—8, XIII, 46.

Concomitant-system: of quintic, X, 342.

Cone: touching six lines, VIII, 401—3; formulæ for potentials of, IX, 266—7.

Cones: through cubic curve in space, III, 219—21; note on cubic, IV, 120—2; and cubic centres, IV, 173—8, 179—81; and cubic curves, V, 284—8; kinds of cubic, V, 401—15; and representation of curve, V, 552; circumscribed sextic, VII, 139; satisfying six conditions, VIII, 99—137; the term asymptotic, XIII, 232; characteristic n and theory of curves in space, XIII, 468—72.

Configurations: algebraic, by Hilbert, VI, 596.

Conformal Representation: IX, 609—11, XI, 442, XII, 104; by function $\arcsin(x+iy)$, X, 290—2; mechanical constructions, X, 406; example, XI, 38; theorem, XI, 78—81; and Schwarzian theory, XI, 169—76; imaginary quantities, XI, 258—60; (*see also* Orthomorphosis, Representation, Transformation).

Congregate: the word, X, 339, 345.

Congruences: in *Ency. Brit.*, XI, 628, 634—5; analytical theory, XIII, 228—30.

Conic: theorem of triangle and line, V, 100—2; theorem of eight points on a, V, 427—30; formula for intersections of line and conic, V, 500—4; four points on, V, 571; defined by five conditions, VII, 546, 552; through three points and with double contact, VII, 554; foci of, VII, 571; and four points, VII, 581, 587; construction, VII, 592; (2, 2) correspondence of points on, VIII, 14—21; and cubic, X, 605—7; Monge's differential equation, XII, 393; focals of quadric surface, XIII, 54.

Conic-node: the term, VI, 360.

Conics, Analytical Theory of: IV, 395—419; relating to single conic, IV, 396—402; ditto with point or line, IV, 402—12; ditto with tangent of conic of double contact, IV, 413; relating to two conics, IV, 416—9.

Conics: general theory of, I, 519—21; inscribed in a quadric surface, I, 557—63; in-and-circumscribed polygon, II, 142—4, IV, 295—9; two dimensional geometry, II, 575—83, 586—92; forms of equations of, III, 86—90; area of, and trilinear equation second degree, III, 143—8; normals of, IV, 74—7; of five-pointic contact of plane curve, IV, 207—39; which touch four lines, IV, 429—31; system having double contact, IV, 456—9, VII, 568; theorem in, IV, 481—3; touching curves, V, 31—2, 552; four inscribed in same conic and passing through same three points, V, 131—2; contact of, V, 552; and rectangular hyperbolas, V, 554; problem, V, 562, 582; tangents of, V, 578; intersection of, V, 582; triangle and, V, 593; and cubic, V, 608; drawing of, VI, 19; locus from two, VI, 27—34; theorem of four which touch same two lines and pass through same four points, VI, 35—9; which touch cuspidal cubic, VI, 249—53; contact of third order with given cuspidal cubic and double contact with given cubic, VI, 253—6; Zeuthen's forms for characteristics of conics which satisfy four conditions, VI, 256—8; correspondence, and those which satisfy given conditions, are at least arbitrary, VI, 268—71; five conditions of contact with a given curve, VI, 272—91; foci of, VI, 517—9, VII, 1—4; determined by five conditions of contact, VII, 40; three, problem and solution, VII, 595; and absolute, VIII, 31—44; theorem of eight points on, VIII, 92—4; cuspidal, of centro-surface, VIII, 352—7; reciprocal of equation, VIII, 522—3; theory of confocal, VIII, 556—7; sets of four points on, X, 569; and lines, X, 602; degenerate forms of curves, XI, 218—20; the term, XI, 460; in *Ency. Brit.*, XI, 561—4; analytical geometrical note on, XII, 424; *f* and *c*, XIII, 11—2; non-existence of special group of points, XIII, 212; the nine-point circle, XIII, 517—9.

Conics, Spherical: theorem relating to, IV, 428; and stereographic projection, V, 106—9; (*see also* Polyzomal curves).

Conics which pass through: four points, III, 136—8; four foci of given conic, IV, 505—9; three given points and touch one line, V, 258—64; two given points and touch two given lines, VI, 43—50; two given points and touch given conic, VI, 245—9.

Conic Torus: the, IX, 519—21.

Conjugate Integrals: Hamiltonian, X, 113—5.

Conjugates: table of, and seminvariants, XIII, 303, 307.

Connected Areas: XI, 7.

Connective: of discriminant, II, 529.

Connective Covariant of two Quantics: defined, II, 515.

Conormal Correspondence: of vicinal surfaces, VIII, 301—8.

Constants: number of, in special equations, XI, 14—6.

Constructive Geometry: VII, 27.

Contacts: problem of, I, 522—31; the term, VII, 546.

Content: Ball on theory of, II, 606.

Continuous Function: the term, XI, 539.

Contour: lines, IV, 108—11, 609; defined, V, 63.

Contracovariants: defined, IV, 329.

Contractible Squarewise: the term, XIII, 179.

Contragredient: the term, IV, 607—8, XIII, 46.

Contraprovectant: defined, II, 514.

Contraprovector: the term, II, 514.

Contrasect: the term, XIII, 485.

Contravariant: the term, II, 320, XIII, 46; of ternary cubic, IV, 325.

Convergence: condition of uniform, XIII, 342—5.

Converging Series: product of, IX, 61.

Convertible Matrices (*see* Matrices).

Convolution: the term, VI, 461—2.

Coordinates: in general theory of geometry, II, 604—6; as functions of parameters, VI, 1—2; polyzomal curves, VI, 498—9, 537; trilinear, XI, 467; Plücker, XI, 467; degenerate curves, XI, 488—9; in *Ency. Brit.*, XI, 546—51, 566—7; illustrative of geometry, XI, 552—6; curvilinear, in *Ency. Brit.*, XI, 637; versus quaternions, XIII, 541—4.

Coordinates of a Line: X, 603, XI, 468.

Coordinates of Points: expressions for, V, 517—8; lines and planes, non-Euclidian geometry, XIII, 489—91.

Coordinates, Six of a Line: VII, 66—98, VIII, 401, X, 287, XII, 42—3, 321; introductory, VII, 66; definition and general notions, VII, 67—9; elementary theorems, VII, 69—73; geometrical considerations, VII, 73—5; linear relations between six coordinates, VII, 75—85; geometrical property of an involution of six lines, VII, 85; four given lines and twofold tractor, VII, 85—6; hyperboloid through three given lines, VII, 86—8; six coordinates defined as absolute magnitudes, VII, 88—9, 96—7; statical and kinematical applications, VII, 89—95; transformation of coordinates, VII, 95—6; formulæ of transformation, VII, 97—8.

Coordinates, Spherical: theory of, and systems of equations, I, 213—23.

Coordinates, Transformation of: I, 123—6, IV, 552—9, XI, 136—42; in *Ency. Brit.*, XI, 558—61, 573—6.

Copfaffian: the term, XIII, 406.

Coriolis, G.: motion of three bodies, IV, 541.

Corpus: Sylvester's theory of the, XIII, 47.

Correspondence: on cubic curves, I, 184, 190; homographic figures, I, 212; theory, VI, 263—91, X, 259—60, XI, 482; in-and-circumscribed triangle, VIII, 222—5; vicinal surfaces, VIII, 301—8; of two variables, IX, 94—5, XII, 104; geometrical representation of imaginary variables, X, 316—23; construction of a, XI, 38; quadric transformation between points and planes, XII, 100—1; of Cartesians, and generators of hyperboloid, XII, 587—9.

Correspondence of Points: V, 542—5, VI, 22, VII, 168—70, XI, 440; two on a curve, VI, 9—13; on a conic, VIII, 14—21; on surfaces, VIII, 200—8; and lines, VIII, 566.

Cos-centre: the word, XIII, 551.

Cotes, R.: central forces, IV, 517, 586.

Cotterill, T.: correspondence of points, VI, 22; problem of envelope and two circles, VII, 573; theorem of Geiser, IX, 506; goniometric problem, X, 295—7.

Counter-barriers: the term, X, 320.

Counter Order: the term, XIII, 268.

Couples: algebraic, I, 128—31.

Cournot, A. A.: motion of a body, IV, 583, 586.

Covariantive Forms and Tables: XI, 277—80; M to W of binary quintic, II, 282—309; asyzygetic, to degree 18, VI, 149—52; 34 concomitants of ternary cubic, XI, 342—56; of binary sextic, XI, 372—6, 377—88; theory of tamisage, XI, 409—10.

Covariants: the term, I, 577, 589, II, 224, IV, 594, 605, X, 340, XIII, 46; determined by differential equations, II, 164—78; theory, II, 164—78; of binary cubic, II, 189, 260—2; binary quadratic, II, 189; binary quartic, II, 190, 262—4; asyzygetic, II, 250; binary quantic, II, 269; of cubic, analogous to invariants of quartic, II, 553; bibliography, II, 598—601; of degree 6, VI, 148—53; of binary cubic, geometrical interpretation, VII, 332—3; the terms asyzygetic and irreducible, VII, 336; theory of number of irreducible, VII, 336—7; also new formulæ for asyzygetic, VII, 337—40; also 23 fundamental, VII, 341—8; Gordan's proof for the number, VII, 348—53; theory founded by Cayley, VIII, xxix—xxx; his work, VIII, xxx—xxxii; as transvectants, VIII, 404—8; connected with an algebraical operation, IX, 537—42; derivatives of three binary quantics, X, 278—86; theorem, X, 430—1; a formula, XI, 122—4; formula and Schwarzian derivative, XI, 184—5; in geometry, XI, 474; Sylvester on, XIII, 47; a hyperdeterminant identity, XIII, 210—11; theory of derivation connected with particular operators, XIII, 329—32; (*see also* Invariants, Linear Transformation, Seminvariants).

Cox, H.: Taylor's theorem, VIII, 493.

Cox, H., Jun.: non-Euclidian geometry, XIII, 481.

Cramer, G.: determinants, I, 63; curve classification, V, 354; transformation of plane curves, VI, 1.

Creedy, C.: tangential of cubic, II, 558; calculations by, III, 361; elliptic motion, IV, 522, 586.

Cremona, L.: on Steiner's quartic surface, V, 423; general theory of correspondence, VI, 22—3; Casey's equation, VI, 66—7; scrolls, VI, 327—8, VII, 245—51; polyzomal curves, VI, 575—6; rational transformation, VII, 189, 200, 207, 222, 253—5, XI, 482, 484; theory of curve and torse, VIII, 72, 76—9, 87—91; geometric transformation, X, 611—2.

Critic Centres (*see* Involution of Cubic Curves).

Criticoids: and invariants, XII, 390; and reciprocants, XIII, 366—7; of Cockle, XIII, 366—7.

Critic Points and Lines: the term, X, 311—5.

Crofton, M. W.: polyzomal curves, VI, 507; Cartesian curves, VII, 582.

Cross-points: the term, X, 317.

Crunodal: the term, V, 402, 551, XI, 228.

Crunode: defined, IV, 181, V, 295, 521, XI, 630.

Crystals: biaxal and ray planes, IX, 107—9.

Cube: axial systems, V, 531—9; automorphic function for, XI, 169, 179—83, 212—6.

Cubic: canonical form of binary, II, 542; equation of differences for, IV, 242, 279; ternary, IV, 325—41; and tables, IV, 333—41; the term, IV, 604; resultant of two binary, V, 289—91; asyzygetic covariants of binary, VII, 338—9; automorphic transformation of binary cubic, XI, 411—6; (*see also* Binary and Ternary Cubics).

Cubic Centres: of lines, V, 73—6.

Cubic Cone: kinds of, V, 401—5, 551, 553; anharmonic property, V, 411—2.

Cubic Curve Classification: V, 354—400, VIII, xxxviii; seven head divisions, V, 355—6; their equations, V, 356—9; thirteen divisions, V, 360—1; notion of group, V, 361; osculating asymptotes, V, 361—3; Newton's classification, V, 364—6, XI, 464; Plücker's, V, 366—8; theory of groups, V, 368—9; groups of hyperbolas △, V, 369—70; hyperbolas △ redundant, V, 370—6; ditto defective, V, 376—88; hyperbolas ⊙, V, 388—9; ditto redundant, V, 389—90; ditto defective, V, 390—1; groups of parabolic hyperbolas, V, 391—4; of central and parabolic hyperbolisms, V, 394; groups of divergent parabolas, V, 395; trident curve and cubical parabola, V, 395; division into species, Newton and Plücker, V, 396—9.

Cubic Curve Memoirs: first, I, 183—9, 586; remarks, I, 190—4, 586; second memoir, II, 381—416, IV, 188; definitions, II, 382—3; theorems relating to conjugate poles, II, 383—5; their proof, II, 385—96; geometrical definition of Quippian, II, 396—7; theorems, II, 397—403; formulæ for intersection of curve and line, II, 404—5; formulæ for satellite point and line, II, 405—9; theorems relating to satellite point, II, 409—12; first polar point of cubic, II, 412—5; recapitulation of geometrical definitions of Pippian, II, 416.

Cubic Curves: tangential of, II, 558—60; cones through, III, 219—21; note on, IV, 120—2; five pointic contact, IV, 231—6; ninth point of intersection of those passing through eight given points, IV, 495—504; twisted, V, 1; sextactic points of plane, V, 233—5; and cones, V, 401—15, 551, 553; inflexions of, V, 493—4, XI, 473; in connexion with quintics and quartics, V, 580; problem, V, 586; derivation of points, VI, 20; intersection of, VI, 20; in pencil of six lines, VI, 105—15, 593—4; nodal, VI, 171—4; foci of circular, VI, 521—2; theory of circular, VI, 526—8; symmetrical circular, VI, 549—50, 550—3; quartic and three, VII, 546; points on, VII, 549; and Cartesian curves, VII, 556; rectangular, VII, 591; mechanical description, VIII, 147—50; residuation in regard to one, IX, 211—4; problem and solution, X, 592—4; equation of, X, 603; and conic, X, 605—7; Abel's theorem applied to, XI, 27—8; degenerate, XI, 220; date of theory, XI, 449; forms and classification, XI, 478—80; circular, XI, 481; systems of, XI, 487; Abel's theorem, XII, 30; elliptic functions, XII, 35—7; as ground-curve in Abel's theorem, XII, 38, 109—216; twisted, on quadric surface, XII, 307—10; notion of, XIII, 79—80; and non-existence of a special group of points, XIII, 212.

Cubic Equations: solution of, II, 542; Tschirnhausen's transformation, IV, 364—7, 377; equation of squared differences, IV, 463—5; Sturmian constants, IV, 473—7; relation between roots, VII, 548; solution by radicals, X, 9; constants of, XI, 556; note on, XII, 421—3; Cardan's solution, XII, 299; on two, XIII, 348—9.

Cubic Forms: letters on, III, 9—12.

Cubic Identity: problem, V, 597.

Cubic Scrolls (*see* Scrolls).

Cubic Seminvariants: generating function, XIII, 306.

Cubic Surfaces Memoir: VI, 359—455, 595—6; Introductory, VI, 359; twenty-three cases, explanations, and tables of singularities, VI, 359—63; determination of number of certain singularities, VI, 364—5; lines and planes of cubic surface, facultative lines, diagrams, VI, 365—6; different kinds of axis, VI, 367; determination of reciprocal equation, VI, 368—70; explanation of sections of memoir, VI, 370—1; equations, $I = 12$, VI, 371—83; $12 - C_2$, VI, 383—90; $12 - B_3$, VI, 391—6; $12 - 2C_2$, VI, 397—402; $12 - B_4$, VI, 403—7; $12 - B_3 - C_2$, VI, 407—11; $12 - B_5$, VI, 411—8; $12 - 3C_2$, VI, 418—

22; $12-2B_3$, VI, 422—6; $12-B_4-C_2$, VI, 426—8; $12-B_6$, VI, 429—30; $12-U_6$, VI, 431—3; $12-B_3-2C_2$, VI, 433—6; $12-B_5-C_2$, VI, 437—9; $12-U_7$, VI, 439—40; $12-4C_2$, VI, 441—2; $12-2B-C_2$, VI, 443—4; $12-B_4-2C_2$, VI, 445—6; $12-B_6-C_2$, VI, 447—8; $12-U_8$, VI, 448—9, 451—5; $12-3B_3$, VI, 449—50; synopsis of foregoing, VI, 450; cubic scrolls, VI, 451.

Cubic Surfaces: triple tangent planes, I, 445—56, 589; skew, V, 90—4; delineation of scrolls, V, 110—2; nodal curve of developable from quartic equation, V, 135—7; theory, V, 138—40; five given lines on, VII, 177—8; double sixers, VII, 316—29; and tetrahedra, VII, 607; Wiener's model with 27 real lines, VIII, 366—84; in *Ency. Brit.*, XI, 633.

Cubic Transformation of Elliptic Functions: III, 266—7, VII, 44—6, X, 46, 58, XII, 518—22, 555, 556—7; geometric illustration, IX, 522—6.

Cubi-Cubic Curves: in space, V, 18—9.

Cubinvariants: of binary quartic, I, 94; of quantic, II, 516; the term, IV, 606; of quadri-quadric function, XIII, 67—8.

Cuboid: potential of, IX, 272, 274—5, 278—80.

Cumulant: the word, IV, 600—1.

Cunningham, A.: on number of terms in a determinant, X, 579—80.

Curtate: the term, XI, 155.

Curvature: lines of, on ellipsoid, I, 36—9; of plane curve at double point, IV, 466—9; of surfaces, IV, 466—9; geodesic, XI, 323—30; (*see also* Curves of Curvature, Orthogonal Surfaces).

Curves: and developables, I, 207—11, 485, 586—7, 589; and two dimensional geometry, II, 569—83; partial branch of, V, 425; reciprocation, V, 505—10; representation by cone and monoid surface, V, 552; nodal, spinode and cuspidal, of cubic surfaces, VI, 450, 595; and space of m dimensions, VI, 456—7; correspondence of two points on, VII, 39; graduation, VII, 426; mechanical description, of, VIII, 138—44, 147—50, 151—5, X, 576; bicyclic chuck for, VIII, 209—11; penultimate forms, VIII, 258—61, 262—3; property of curve and torse, VIII, 520—1; coordinates and equations, X, 546; degenerate forms, XI, 218—20, 487—9; abstract geometry, XI, 441—2; in *Ency. Brit.*, XI, 460—89, 572—3, 579—80; and theory of equations, XI, 501; and function, XI, 540—1; and solid geometry, XI, 569; quadrature of, XI, 641—2; minimal surfaces, XIII, 41; orthotomic, of a system of lines in a plane, XIII, 346—7; (*see also* Correspondence, Cubic Curves, Nodal Curves, Polyzomal Curves).

Curves, Algebraic: I, 46—54, 584.

Curves, Bicursal: VIII, 181—7.

Curves, Classification of: V, 613—7; (*see also* Cubic Curve Classification).

Curves, Excubo-quartic: V, 282.

Curves in Space: analytical representation, IV, 446—55, 490—5, 616—8, VII, 66, XI, 9—13; defined by conoid and monoid surfaces, V, 7—20, 552, 553, 613; quartic, V, 11—5; quintic, V, 15—6, 24—30, 552, 553, 613; quadri-cubic, V, 16; quadri-quartic, V, 17; cubi-cubic, V, 18—9; Halphen's characteristic n in theory of, XIII, 468—72.

Curves, Intersections of: I, 25—7, 583, XII, 500—4; real, IX, 21.

Curves of Curvature: near umbilicus, VII, 330—1; on surfaces, VIII, 97—8, 145—6, 264—8; *Ency. Brit.*, XI, 628, 635—6; wave surface, XII, 249; surfaces with spherical, XII, 601—38.

Curves of Striction: I, 234.

Curves, Opposite: V, 468.

Curves, Parallel: envelopes and surfaces, IV, 123—33, 152—7, 158—65; and evolutes, VIII, 31—5; theory of, X, 260; the critic, in solar eclipses, X, 311—5.

Curves, Pedal: V, 113—4.

Curves, Penultimate Quartic: VIII, 526—8.

Curves, Penumbral: geometrical theory of projection, VII, 483, 488—9, 489—92.

Curves, Plane: double tangents of, IV, 186—206; conic of five-pointic contact of, IV, 207—39;

curvature at double point, IV, 466—9; higher singularities, V, 424—6, 520—8, 619; correspondence of points on, V, 542—5; transformation, VI, 1—8, 593, VIII, 387; notion of, of given order, XIII, 79—80.

Curves, Plane, sextactic points of: V, 221—57, 618—9; condition for point, V, 222—5; notations and remarks, V, 225—6; first transformation, V, 226; second, V, 227—8; third, V, 228—9; fourth and final form, V, 229—33; application to cubic, V, 233—5; proof of identities, V, 235—7; Jacobian formula, V, 237—8; proofs of equations and identities, V, 239—47; appendix, V, 247.

Curves, Poloid: IV, 571.

Curves, Rhizic: IX, 34.

Curves, Serpoloid: IV, 571.

Curves, Sextic: VII, 256—7, VIII, 138—44, X, 612.

Curves, Symmetric: I, 473.

Curves, Theory of: and elimination, I, 337—51, V, 162—7, 416—20; evolution, XI, 449—51, XII, 102—3, 290—1.

Curves, Theory of, and Torse: VIII, 72—91; explanations and notation, VIII, 72—4; Plücker-Cayley equations, VIII, 74, 75—6, 80—1; Salmon-Cremona equations, VIII, 74, 76—9, 87—91; geometrical theory of foregoing relations, VIII, 79—80; tables, VIII, 81—4; nodal curve x, VIII, 84—7.

Curves, Three-bar: IX, 551—80, 585.

Curves, Transformations of: I, 471—5, 476—80; scalene, IX, 527—34.

Curves, Triangular: VII, 59.

Curves, Twisted Quartic: XII, 428—31.

Curves which satisfy given conditions: VI, 191—262, 594, VII, 40; Introductory, VI, 191; previous memoirs, VI, 191—2; quasi-geometrical representation of conditions, VI, 193—200; Chasles' and Zeuthen's researches, VI, 200—26; extensions of de Jonquières, VI, 226—42; form of equation of curves of a series of given index, VI, 242—3; line-pairs which pass through three given points and touch a given conic, VI, 244; conics which pass through two given points and touch given conic, VI, 245—9; conics which touch cuspidal cubic, VI, 249—53; conics which have contact of third order with given cuspidal cubic and double contact with given cubic, VI, 253—6; Zeuthen's forms for characteristics of conics which satisfy four conditions, VI, 256—8; question from de Jonquières' formula, VI, 258—62; the principle of correspondence, VI, 263—91; (introductory, VI, 263—4; correspondence of two points on a curve, VI, 264—8; application to conics which satisfy given conditions, one at least arbitrary, VI, 268—71; five conditions of contact with a given curve, VI, 272—91).

Curve Tracing: Cayley's liking for, VIII, xxxix; mechanism, VIII, 179—80, XIII, 515—6; importance, XI, 461; order of, XI, 461.

Curvilinear Coordinates: XI, 330, XII, 1—18; surfaces divisible into squares, VIII, 146; geodesic lines, VIII, 156—67; curves of curvature, VIII, 264—8; orthogonal surfaces, VIII, 269—91.

Cusp: of Cartesian at circular points at infinity, I, 589; synonymous with spinode, II, 28, IV, 22, 27; of second kind or node-cusp, V, 265—6, 618; order of plexus for, V, 309—12; the term, XI, 468.

Cuspidal: defined, V, 403, 551, VII, 244.

Cuspidal Conic: of centro-surface, VIII, 352—7.

Cuspidal Cubic: VII, 561.

Cuspidal Curves: and cubic surfaces, VI, 450; (see also Cubic Surfaces, Surfaces).

Cuspidal Isochronic: the term, VII, 473.

Cuspidal Locus: in singular solutions, VIII, 533.

Cyc: the abbreviation in groups, XIII, 119.

Cyclide: of Dupin, V, 467, XII, 615; the term, VII, 246, VIII, 262, IX, 64—5; and anchor ring, IX, 18; on, IX, 64—78; the parabolic, IX, 73—8; in *Ency. Brit.*, XI, 634.

Cycloid: the term, XI, 447.

C. XIV.   12

Cyclotomy: (Kreistheilung), XI, 58, 86.

Cylinder: in *Ency. Brit.*, XI, 572—3; Archimedes' theorem for surface, XII, 56—7.

d'Alembert, J. le R.: rotation of solid body, IV, 567, 586; geometric paradox, XII, 305—6.

Dandelin, G. P.: theory of, and on caustics, II, 339—40; wave surface, IV, 433—4.

Darboux, G.: powers of circles and spheres, I, 581; the torus, VII, 247; quartic surfaces, VIII, 262; cyclide, IX, 64; continuous function, XI, 539; curves of curvature, XII, 615; quartics, XIII, 13.

Davis, W. Barrett: calculations by, III, 361, IV, 376; quantic covariants, VII, 335; Sohnke's modular equations, IX, 543.

Déblais: theory of, XI, 417—20, 449, 587.

de Bruno, Faà: invariants of degree 12 belonging to quintic, II, 314; symmetric functions, II, 602; canonical forms, IV, 52; elimination, IV, 608.

Decadianome: the term, VII, 134; and symmetroid, VII, 256.

Decapitation: and seminvariants, XII, 248—50.

Decomposition: linear differential equations and theory of, XII, 402, 403—7.

Dedekind, J. W. R.: probabilities, II, 594—8, V, 80; modular function, XIII, 338—41.

Def: the abbreviation in groups, XIII, 120.

Deficiency: and genus of curve, V, 467, 517, 619; of curves, V, 618, VIII, 391, XI, 450; the term, VI, 2; and transformation, VI, 3, XI, 482—5; the term applied to surface, VI, 356; of certain surfaces, VIII, 394—7, XI, 230; surfaces of negative, VIII, 397; of sextic curve, IX, 504—7; of curve and Abelian integrals, XI, 30—6.

Definite Integrals: with complex variables, I, 181—2, 310; differentiation and evaluation, I, 267—72, 587; on a, IV, 28—9; note on Glaisher's paper, VIII, 1; note on two, IX, 56—63; (*see also* Attractions, Potentials, Prepotentials).

Definitions (*see* the word desired).

Deformation: the term, I, 234; of hyperboloid, XI, 66—7; of skew surfaces, XI, 317—22, 331; (*see also* Surfaces).

Degeneracy: of surfaces, V, 98—9; of scrolls, V, 201—3; of curves, XI, 218—20.

Degen's Mathematical Tables: IV, 40, IX, 478—9, X, 586; report of British Association Committee, XIII, 430—67.

Deg-order: the term, X, 339.

Degrees: of quantics defined, II, 221.

Degrees, Honorary: conferred on Cayley, VIII, xx—xxi; conferred on Sylvester, XIII, 43.

de Jonquières, E.: cubic curves, I, 586, IV, 496; on curves, IV, 454; curves which satisfy given conditions, VI, 191, 192, VII, 41—3; form of equation of curves of a series of given index, VI, 242—3; question from formula, VI, 258—62; points on cubic curve, VII, 550, 553; correspondence, XI, 486.

de la Goupillière, H.: inertia, IV, 566.

de la Gournerie, M.: tetrahedral scrolls, VII, 48—53; scrolls, VII, 54; quartic and quintic surfaces, VII, 246, 247, 251; torus, VIII, 25; octic surfaces, X, 81.

Delambre, J. B. J.: Tables du Soleil, III, 474.

Delaunay, C. E.: lunar theory, VII, 357, 372, 376, 528—33, 534, IX, 180, XIII, 206—7.

de Morgan, A.: root in every algebraic equation, IV, 116—9; root limitation, IX, 39; series, XI, 623, 627; and Sylvester, XIII, 45.

Denumerant: illustrated, IV, 169—70.

Denumerate: defined, IV, 241.

Departure Point: in lunar theory, III, 19, 270, 295.

Derivation: of points of cubics, VI, 20; and Übereinanderschiebung, VII, 348; and seminvariants, XIII, 362—5.

Derivational Function: the term, I, 63.

Derivations: extension of Arbogast's method, II, 257, IV, 265—71, 272—5, 609, XI, 55; binomial theorem and factorials, VIII, 463—73.

Derivatives: and hyperdeterminants, I, 95; of point on cubic, IV, 231; of three binary quantics, X, 278—86; [and covariants, X, 340, 377—94; of a function, X, 590—2; in binary forms, XI, 272; (*see also* Schwarz).

de St Laurent, T.: caustic by reflexion, I, 273—5.

Desboves, A.: planetary perturbation, III, 185, 203; problem of two centres, IV, 532, 586.

Descartes, R.: ovals of, and transformation of curves, I, 478, 479—80, 589; oval of, III, 66; formulæ in *Epistolæ*, IV, 512; geometry of, XI, 437; (*see also* Cartesians).

Determinants: applied to distances of points, I, 1—4, 581, IV, 510—2; Pascal's theorem, I, 43—5; the term, I, 63; theory of, I, 63—79; theory of linear transformations, I, 80—94, 584; of vis viva, I, 284; note on hyperdeterminants, I, 352—5, 588; geometrical reciprocity, I, 377—82; "skew" and "symmetric," I, 410—3; history, I, 581; multiplication, I, 581, XI, 495; value of certain, III, 120—3, IV, 460—2; the term, IV, 594, 596—9; and Pfaffian, IV, 600; development of, V, 45—9; tables of binary cubic forms for negative, VIII, 51—64; Smith's Prize dissertation, VIII, 551—5; symmetrical, IX, 185—90, X, 579; notation, X, 95—7; theorem in, X, 265—6; in *Ency. Brit.*, XI, 490—7; decomposition, XI, 495—6; theory of numbers, XI, 604—9; (*see also* Hyperdeterminants, Skew Determinants).

Determinator: defined, II, 59.

Determinirende: (indicial), and differential equations, XII, 398, 401, 453.

Developables: and curves, I, 207—11, 586—7; the term, I, 486, XI, 573; from two quadrics, I, 486—95; from quintic curve, I, 500—6; planar, I, 505; from quartic, V, 135—7; prohessians, V, 267—83; quartics, V, 268—71; general theory, V, 271—2; special quintic, V, 272—8; special sextic, V, 279—83; reciprocation of quartic developable, V, 505—10; a special sextic, V, 511—9; sextic, and sextic surfaces, VI, 87—100; focals of a quadric surface, XIII, 51—4; (*see also* Torse).

Development: of factorial, II, 98—101; coefficients in powers of $(1 + n^2 x)^{m/n}$, XIII, 354—7.

Dew-Smith, A. G.: portrait of Cayley, XI (frontispiece).

Diagonals: and partitions of a polygon, XIII, 93—113.

Diagrams: the term, VII, 405; of planet's orbit from three observations, 5 plates, VII, to face 478; solar eclipse, VII, to face 492; geodesic lines on ellipsoid, VII, 510; coloured, representing groups, X, 328—30; transformation of elliptic functions, XI, 26; seminvariants, and solution by square-, XIII, 288—98; (*see also* Tables).

Diameter: as used by Newton, V, 362.

Diametral planes (*see* Planes).

Dianome: the term, VII, 133, 148; (*see also* Quartic Surfaces).

Diaphoric: the term, XI, 156.

Dickinson, L.: portraits of Cayley, VI (frontispiece), VII (frontispiece), VIII, xx.

Differences: equation of squared, for cubic, IV, 463—5; relation between certain products of, X, 293—4; on a functional equation, X, 298—306; (*see also* Equation of Differences).

Differential Equation Memoir: X, 93—133; introductory, X, 93—4, 94—5; notations, X, 95—7; dependence of functions, X, 97; general differential system, X, 98—102; the Multiplier, X, 102—5; Pfaffian theorem, X, 106; Hamiltonian system, derived from general system, X, 106—7; Poisson-Jacobi theorem, an identity in regard to functions (H, Θ), X, 108—9; peculiar to Hamiltonian system, X, 110—3; conjugate integrals of Hamiltonian system, X, 113—5; Hamiltonian system—the function V, X, 115—8; partial differential equation $H = $ constant, X, 119—25; examples, X, 125—32; partial differential equation containing the dependent variable, reduction to standard form, X, 132—3.

Differential Equations: and lines of curvature of ellipsoid, I, 36—9; dynamical, I, 276—84; Jacobi's system of, ɪ, 366—9 ; Jacobi on theory, ɪɪɪ, 174 ; theorem of Jacobi on Pfaff's problem, ɪᴠ, 359—63; singular solutions, ɪᴠ, 426—7; transformation, ɪᴠ, 574, ᴠ, 78—9 ; umbilici, ᴠ, 115—30 ; solution when algebraical, ᴠɪɪ, 5—7 ; supposed new integration, ᴠɪɪ, 36 ; note on one, ᴠɪɪ, 354—6 ; pair in lunar theory, ᴠɪɪ, 535—6, 537—40 ; integration by series, ᴠɪɪɪ, 458—62 ; Euler's, ɪx, 592—608 ; and theory of elliptic functions, x, 20 ; and sides of quadrangle, x, 33—5 ; theory of partial, x, 134—8 ; elliptic and single theta functions, x, 422—9 ; hypergeometric series, xɪ, 17—25 ; Abel's theorem, xɪ, 27—8 ; new formulæ for integration of Euler's equation, xɪ, 68—9 ; mathematics and physics, xɪ, 449 ; connected with elliptic functions, xɪɪ, 30—2 ; Briot and Bouquet's theory, xɪɪ, 432—41 ; of circular functions, xɪɪ, 580 ; a diophantine relation, xɪɪ, 596—600 ; and construction of Milner's lamp, xɪɪɪ, 3—5 ; Kummer's, of third order, xɪɪɪ, 69—73 ; on a partial, xɪɪɪ, 358—61 ; Richelot's integral of Euler's, xɪɪɪ, 525—9 ; (*see also* Partial Differential Equations, Riccati, Schwarz, Singular Solutions).

Differential Equations, Linear : invariants of one, xɪɪ, 390—3 ; general theory, xɪɪ, 394—403, 444—52, 453—6 ; theory of decomposition, xɪɪ, 403—7.

Differential Equations of First Order : theory of singular solutions, ᴠɪɪɪ, 529—34, x, 19—24.

Differential Invariants (*see* Invariants).

Differential Operators : ᴠɪɪ, 8.

Differential Relations : of double theta-functions, x, 559—65.

Differentiation : evaluation of definite integrals, ɪ, 267—72, 587 ; formulæ for, ɪᴠ, 135—49 ; fractional, xɪ, 235—6.

Dimensions in Geometry (*see* Geometry).

Dimidiate : the term, xɪɪɪ, 119.

Dimidiation : the term, xɪɪɪ, 122.

Diophantine Differential Relation : xɪɪ, 596—600.

Diptich : the term, xɪɪ, 596.

Dirichlet, G. L. (*see* Lejeune-Dirichlet).

Director, Nodal (*see* Nodal Director).

Directrix : and scrolls, ᴠɪɪ, 60 ; and the absolute, xɪɪɪ, 481—9, 501 ; kinematics of a plane, xɪɪɪ, 505—6.

Discriminant : and invariant, ɪ, 584 ; defined, ɪɪ, 176, ɪᴠ, 603, ᴠɪ, 466—7 ; of quantics, ɪɪ, 320 ; the sign □, ɪɪ, 528 ; special, connected with curve, ᴠ, 163 ; of quintic, problem, ᴠ, 592 ; of binary quantic, ᴠɪɪ, 303, ɪx, 16—7 ; example of a special, ᴠɪɪɪ, 46—7 ; (*see also* Quantics).

Discriminant Locus : the term, ᴠɪ, 198.

Displacement : the term in Abel's theorem, xɪɪ, 110, 157—62.

Distance : general theory of, ɪɪ, 561, 583—92, 604—6, ᴠ, 550 ; notion of, in analytical geometry, ᴠ, 550 ; the term, ᴠɪ, 497 ; angular, of two planets, ᴠɪɪ, 377—9 ; Cayley and Klein on theory, ᴠɪɪɪ, xxxvi—vii ; general notion, ᴠɪɪɪ, 31 ; Euclidian geometry, xɪ, 435—7 ; non-Euclidian geometry, xɪɪɪ, 480—504 ; (*see also* Points).

Distribution of Electricity : on spherical surfaces, ɪᴠ, 92—8, 99—107, xɪ, 1—6.

Distributively : the term, ᴠɪ, 459.

Disturbing Function : in lunar theory, ɪɪɪ, 293—308, 319—43 ; in rotation of solid body, ɪɪɪ, 486.

Divisors : tables of, ɪx, 462—70.

Dodecahedron : construction, ɪᴠ, 82—3 ; axial systems, ᴠ, 531—9 ; as regular solid, x, 270—3 ; automorphic function for, xɪ, 169, 179—83, 184, 212—6.

Donkin, W. F. : expansions in multiple sines, ɪ, 583 ; differential equations, dynamical, ɪɪɪ, 191—7, 203—4 344 ; transformation of trigonometric series, ɪɪɪ, 567 ; attractions, ɪɪɪ, 567 ; a definite integral, ɪᴠ, 29 formulæ for differentiation, ɪᴠ, 135—49 ; central forces problem, ɪᴠ, 521 ; spherical pendulum, ɪᴠ, 534, 536, 586 ; dynamical problems, ɪᴠ, 547, 586 ; elimination of nodes in three bodies, ɪᴠ, 551, 586 ; rotation of solid body, ɪᴠ, 578, 586.

Dostor, G. : polyhedra, ɪᴠ, 609.

Dots: notation for lines and planes of cubic surfaces, vi, 365—6, 373—449; and seminvariants, xiii, 267.

Double Algebra: xii, 465.

Double Contact: conics having with each other, iv, 456—9.

Double Point: the term, vi, 1; on ground-curve, xii, 110, 129.

Double Pyramid (*see* Polygons).

Double-Sixer: and cubic surfaces, vi, 372, vii, 316—29; construction, viii, 366—84.

Double Tangents (*see* Bitangents).

Double Theta Functions: x, 155—6, 166—79, 180, 422—9, 474—5, 497, 565; in connexion with 16-nodal quartic surface, x, 157—65; memoir on, x, 184—213; (Part I, preliminary investigations, x, 184—9; Part II, the double theta functions, x, 189—213); addition of, x, 455—62; evolution, xi, 454; transformation, xii, 358—89; (*see also* Theta functions).

Doubly Infinite Products: i, 120—2, 132—5, 136—55, 156—82, 585, 586, x, 492—4, xi, 46, xii, 50—5; and doubly periodic functions, ii, 150—63; and definite integrals, ix, 60; transformation of, x, 494—7.

Doubly Periodic Functions: i, 156—82; and doubly infinite products, ii, 150—63; and definite integrals, ix, 61; the term, xi, 530.

Drawing: geometrical, vi, 19; of quartic curves mechanically, viii, 151—5; curves generally, viii, 179—80; (*see also* Representation).

Droop, H. R.: isochronism of circular hodograph, iii, 265; central forces problem, iv, 520, 587.

Duality: in geometry, ii, 561—2, 568, xi, 450, 467.

Du Bois-Reymond, P. L.: uniform convergence, xiii, 343.

Dumas, W.: spherical pendulum, iv, 534, 587.

Dupin, C.: cyclide of, v, 467, ix, 64, xii, 615; quartic and quintic surfaces, vii, 246; theorem of, viii, 264—8, 562, ix, 84—9.

Duplication of Groups: x, 149—52.

Durège, H.: Landen's theorem, xi, 339.

Durfee, W. P.: symmetric functions, ii, 602—3.

Dynamics: differential equations of, i, 276—84; a class of problems, iv, 7—11; similarity of two dynamical systems, viii, 558—63; Lagrange's general equation in, ix, 110—2, 198—200; general equations in, ix, 215—7; and time, xi, 444; transformation of coordinates, xi, 575.

Dynamics, Recent Progress in Theoretical: iii, 156—204, iv, 514; introduction, iii, 156—7; Lagrange, *Mécanique Analytique*, iii, 157—8, 201, 202; Lagrange, equations of motion, iii, 158, 200; lunar theory, iii, 158—9; Poisson, planetary theory, iii, 159, 201; Laplace's theory, iii, 159, 201; Lagrange's planetary theory, iii, 159—61, 162—3, 201; Lagrange, variation of arbitrary constants in mechanical problems, iii, 161—5, 200; also Poisson, iii, 163—5, 200, 201, 202; Cauchy, differential equations, iii, 166; Hamiltonian method of dynamics, iii, 166—74, 200, 202; its relations to Lagrange's, iii, 171—3, 200; and Poisson, iii, 173—4, 200; Jacobi, calculus of variations and differential equations, iii, 174—82, 200, 202; *De Motu Puncti Singularis*, iii, 182—3, 202; problem of three bodies, iii, 183; Jacobi, *Theoria Novi Multiplicatoris*, iii, 183—5; Jacobi, theory of ideal coordinates, iii, 185; Liouville, equations of motion, iii, 185; Desboves, planetary perturbation, iii, 185, 203; Serret, integration of differential equations, iii, 185—6, 203; Sturm, integration of dynamical equations, iii, 186, 203; Ostrogradsky, dynamical equations, iii, 186, 203; Brassinne, differential equations, iii, 186—7, 203; Bertrand, integrals to mechanical problems, iii, 187, 203; and integration of differential equations, iii, 188—9, 203; and *Mécanique Analytique*, iii, 189—90, 203; Brioschi, *Sulla Variazione*, and *Teorema di Meccanica*, iii, 190, 203; Liouville, integration of differential equations, iii, 191—2, 203; Donkin, dynamical differential equations, iii, 191—7, 203—4; Bour, integration of differential equations of analytical mechanics, iii, 197—8, 204; Liouville on Bour's memoir, iii, 199, 204; Brioschi, *Degli Integrali di un Problema di Dinamica*, iii, 199—200, 203; Bertrand, integrals of several mechanical problems, iii, 200, 203; summary, iii, 200.

Dynamics, Report on Progress of Solution of Certain Problems: iv, 513—93; introductory, iv,

513—5; rectilinear motion, IV, 515—6; central forces, IV, 516—26; elliptic motion, IV, 521—4; problem of two centres, IV, 524—32; spherical pendulum, IV, 532—4; motion as affected by the Earth and relative motion generally, IV, 534—7; motion of single particle, IV, 537—8; motion of three mutually attracting bodies in a right line, IV, 538—40; motion of three bodies, IV, 540—1; motion in resisting medium, IV, 541; integration of equations of motion, IV, 542—6; memoirs by Jacobi, Bertrand and Donkin, IV, 546—7; problem of three bodies, IV, 548—52; transformation of coordinates, IV, 552—9; principal axes and moments of inertia, IV, 559—66; rotation of solid body, IV, 566—80; kinematics of solid body, IV, 580—2; rotation round fixed point, IV, 582—3; other cases of the motion of a solid body, IV, 583—4.

Earth, The : rotation of, III, 485, IV, 534—7; (*see also* Gravity).

Eclipses (*see* Solar Eclipses).

Edge : defined, V, 63.

Eindeutig : (Uniform function), XII, 433.

Eisenstein, F. G.: linear transformations, I, 90, 101, 111, 113—6, 585; hyperdeterminants, I, 353, II, 598—601; elliptic functions, I, 586; cubic forms, III, 9; quadratic residues, III, 39—43; finite differences, IV, 263; mathematical tables, IX, 492—3; "development of an idea of," X, 58—9; development of $(1 + n^2x)^{m/n}$, XIII, 357.

Elastic Strings : problem with, III, 78—9.

Electricity : distribution on spherical surfaces, IV, 92—8, 99—107, X, 299, 307, XI, 1—6.

Elements : Jacobi's canonical, III, 77; of arc, X, 235—7; a reduction to elliptic integrals, X, 239—42.

Eliminant: the term, IV, 597; of two quantics, XI, 100—2.

Elimination: and theory of curves, I, 337—51, V, 162—7, 416—20; from connected equations, I, 370—4; and linear transformations, I, 457; theorem of Schläfli, II, 181—4; a result of, III, 214—5; general theorem, IV, 1—4; Bezout's method, IV, 38—9, V, 555—6; of nodes in three bodies, IV, 551; the term, IV, 594; text-books on, IV, 608; note on, V, 157—9; problem, VI, 40—2, VIII, 22—4; the resultant of a system of two equations, VI, 292—9; theorem, IX, 43—4; formula of, XI, 100—2; theory of equations, XI, 490; a problem of Sylvester's, XIII, 545—7.

Ellipse : curves parallel to, IV, 123—33, 152—7; and circle, V, 561; foci of, V, 586; and quadrilateral, V, 604; circles of curvature, VII, 555, XIII, 537; potential, IX, 281—90; negative pedals, X, 576; cubic curves, XI, 478; in *Ency. Brit.*, XI, 561—4; focals of quadric surface, XIII, 54; and epitrochoid, XIII, 82—7; orthomorphosis into a circle, XIII, 188—9, 422—4.

Ellipsoid, Attraction of: I, 388—91, 432—44, 582, VII, 380—3, XI, 448; Jacobi's method, I, 511—18; Gauss's method, III, 25—8, 149—53; Laplace's method, III, 53—65; Rodrigues' method, III, 149—53; theory of, III, 154—5; and terminated straight line, VII, 31—3.

Ellipsoid, Centro-Surface of, memoir : VIII, 316—65; introductory, VIII, 316—7; the ellipsoid, VIII, 317—20; sequential and concomitant centro-curves, VIII, 320; expressions for coordinates of point on centro-surface, VIII, 320—4; discussion by means of equations, principal sections, &c., VIII, 324—30; generation of surface considered geometrically, VIII, 330—1; nodal curve, VIII, 332—52; eight cuspidal conics, VIII, 352—7; centro-surface as envelope of quadric, VIII, 357—8; another generation of centro-surface, VIII, 359—61; a third generation of centro-surface, VIII, 361—2; reciprocal surface, VIII, 363; delineation of centro-surface for particular case, VIII, 363—5.

Ellipsoid, Geodesic Lines on, memoir : VII, 493—510; introductory, VII, 493—4; course of the lines, VII, 494—5; lines through an umbilicus, VII, 495—501; formulæ, VII, 501—3; umbilicar geodesics, VII, 503; tables, VII, 504—6; projection on umbilicar plane, VII, 507; elliptic function formulæ, VII, 507—10; diagram, VII, to face 510.

Ellipsoids : lines of curvature, I, 36—9; surface parallel to, IV, 123—33, 158—65, X, 575; the momental, IV, 560; of gyration, IV, 560; central, IV, 564; projection, V, 487—8; geodesic lines on, VII, 34—5; centro-surface and sextic torse, VII, 113—4; centro-surface, VII, 130—2; geodesic lines, VIII, 174—8;

attraction of ellipsoidal shell on exterior point, IX, 302—11; negative pedals, X, 576; in *Ency. Brit.*, XI, 576—9.

Elliptic Coordinates: equation of wave surface in, XI, 71—2.

Elliptic Functions: Bronwin on, I, 118, 119; of Jacobi, I, 127, 507, 586; integral calculus, I, 383; multiplication of, I, 534—9, 568—76, 589; addition of, I, 540—9, 589, XII, 294—8; connected with theory of numbers, II, 48; system of modular symbols, IV, 484—9; Weierstrass, V, 33—7; treatise by Cayley, VIII, xviii, xxviii, XIII, 560; a general differential equation, IX, 592—608; a differential equation in theory of, X, 24; and integration, X, 25—7; torse depending on, X, 73—8; reduction of Abelian integrals to, X, 214—22; and single theta functions, X, 422—9, 463, 472; certain algebraic identities, XI, 130—1; evolution of, XI, 451—5; and quartic function, XI, 483; kinds of, XI, 529; symmetrical differential equation and, XII, 30—2; solution of $x^3 + y^3 - 1 = 0$, XII, 35—7; Weierstrassian and Jacobian compared, XII, 425—7; Kiepert's *L*-equations, XII, 490—2; graphical representation, XIII, 9—19; and sextic resolvent equations, XIII, 473—9; theta and omega functions, XIII, 558—9; (*see also* Gudermannian, Theta functions).

Elliptic Functions Formulæ: Serret's, III, 3; for geodesic lines on special ellipsoid, VII, 507—10; one, XI, 65, XII, 292--3; connexion of certain, XI, 250—1; geometrical interpretation of certain, XII, 107.

Elliptic Functions, Inverse: I, 136—55, 156—82, 586; and definite integrals, II, 3.

Elliptic Functions, Memoir on Transformation of: IX, 113—75; introductory, IX, 113—4; the general problem, IX, 114—7; Ωk modular equations, IX, 117—8; equation-systems, IX, 119—20; Ωk form, IX, 121—6; modular equation, IX, 126—37; tables, IX, 128—35, 163; multiplier equation, IX, 138—40; multiplier as rational function of u, v, IX, 140—4; multiplication of elliptic functions, IX, 144—7; transformations, IX, 147—55; general theory of q-transcendents, IX, 155—69; four forms of modular equation and curves represented thereby, IX, 169—75.

Elliptic Functions, Theorems in: XI, 73—7; Landen's, XI, 337—9, 584; Hermite's *H*-product, XII, 584—6.

Elliptic Functions, Theory of: I, 290—300, 364—5, 402—4, 587, 589; and quadri-quadric curve, XII, 321—5.

Elliptic Functions, Transformation of: I, 120—2, 132—5, 585, IX, 543, X, 333—8, 611, XI, 26, XII, 416—7, 505—34, 535—55, XIII, 29—32; cubic, III, 266—7, VII, 44—6, 244—5, 253—6, XII, 46, 556—7, XIII, 64—5; special quartic, IX, 103—6; geometric illustration of cubic, IX, 522—6; orthomorphosis, XIII, 191—205.

Elliptic Integrals: reduction of $\dfrac{du}{\sqrt{U}}$, I, 224—7; transformation, I, 508—10, IX, 618—21; geometrical representation, II, 53—6, 113—7; discussion, II, 93—5; and covariants, II, 189—92; transformation formulæ, IV, 60—9, 609; expression for second kind of, X, 25—7; note on theory, X, 139—42; some formulæ in, X, 143—8; of third kind, X, 489—92; problem, X, 614; theta functions, XI, 41—6; note on, XI, 64; reduction of an integral to, XI, 270—1; of third kind, formulæ, XI, 340—1.

Elliptic Motion: expansion of true anomaly, III, 139—42, 567; trilinear equation of second degree, III, 143—8; theory of, III, 216—8; tables of functions in theory, III, 360—474, VII, 516; Lambert's theorem, III, 562—5, VII, 387—9; and dynamical progress, IV, 521—4; a theorem, IX, 191—3; and body let fall at equator, IX, 241—3.

Elliptic Motion, Disturbed: memoirs on, III, 270—92, 344—59, 505—15.

Elliptic Space: and non-Euclidian geometry, XIII, 481.

Elliptic-Transcendent Identity: VIII, 564.

Ellis, R. L.: orthogonal surfaces, VIII, 272; differential equations, VIII, 458; Dupin's theorem, IX, 88.

Emanants: of quantics, II, 321; theory, II, 518; Bezoutoidal, II, 525, 526; the term, IV, 604, XIII, 46.

Emanation: theory of, II, 321.

Encke, J. F.: *über die speciellen Störungen*, III, 179—80; fluctuating functions, IX, 19; roots of numerical equations, X, 5.

Encyclopædia Britannica, articles from on:—Curve, XI, 460—89; Equation, XI, 490—521; Function, XI, 522—42; Galois, XI, 543; Gauss, XI, 544—5; Geometry (analytical), XI, 546—82; Landen, XI, 583—4; Locus, XI, 585; Monge, XI, 586—8; Partition of numbers, XI, 589—91; Theory of numbers, XI, 592—616; Series, XI, 617—27; Surface, XI, 628—39; John Wallis, XI, 640—3.

Endecadic Transformation : in elliptic functions, IX, 152—5.

Endoscopic : the term, I, 588.

Ennead : the term, VII, 256, VIII, 566.

Enneadianome : the term, VII, 134.

Enneagon : in-and-circumscribed, IV, 298—303.

Envelopes : developable of two equations, I, 486; parallel curves and surfaces, IV, 123—33, 152—7, 158—65; defined, IV, 458, VI, 467; of circle, V, 610, VII, 591; and locus in regard to triangle, VI, 72—82; depending on two circles, problem and solution, VII, 573; of plane curve, VII, 606; of a certain quadric surface, VIII, 48—50; locus in singular solutions, VIII, 533; problem of, VIII, 491—2; of family of quadrics, X, 589; theory, XI, 50—1; of variable curves, XI, 475—6.

Epicycloid : and caustic, II, 345.

Epispheric integrals : Gauss-Jacobi theory, IX, 410—7.

Epitrochoid : XIII, 81—7.

Equal : applied to circles, VIII, 31.

Equality : among roots of an equation, II, 465—70, VI, 300—12; idea of, XI, 431.

Equation of Differences : for equation of any degree, IV, 150—1; for equation of any order, IV, 240—61; tables, IV, 246—56; of all but one, of roots of given equation, IV, 276—91; and quintic equation, IV, 309—24, 609—16; and cubic equation, IV, 463—5.

Equation, Pellian (*see* Pellian Equation).

Equations : systems of spherical coordinates, I, 213—23; with quantics, defined, II, 221; auxiliary for quintics, IV, 309—24; determination of reciprocal, with cubic surfaces, VI, 368—70; the term, VI, 466; solubility by radicals, VII, 13—4; system of, problem and solution, VII, 578, 581, X, 601; transformation, IX, 42, 48—51; on a functional equation, X, 298—306; Cassinian, problem, X, 608; Jacobian sextic, XI, 389—401, XII, 493—9; equal roots of, XI, 405—7; of curves, XI, 462—4; of Plücker, XI, 469—73, XIII, 536; in *Ency. Brit.*, XI, 490—521, (introductory, XI, 490; determinant, XI, 490—7; imaginary, XI, 502—6); of right line and circle, XI, 558—61; of conics, XI, 563; seminvariants, XII, 19—21; fundamental, and deformation of surfaces, XI, 331; note on system of, XII, 48—9; for three circles which cut each other at given angles, XII, 559—61, 564—70; anharmonic ratio, XII, 578—9; hydrodynamical, XIII, 6—8; Sylvester on ternary cubic-form, XIII, 47; on soluble quintic, XIII, 88—92; Waring's formula for sum of mth powers of roots of, XIII, 213—6; sextic resolvent of Jacobi and Kronecker, XIII, 473—9.

Equations, Algebraic : rationalization, II, 40—4; theory, II, 124; theorem that every one has a root, IV, 116—9; system of, IV, 171—2, VIII, 29—30; in *Ency. Brit.*, XI, 506—21; Anglin's formula for successive powers of the root of, XII, 33—4; roots of one, XIII, 33—7.

Equations, Cubic (*see* Cubic equations).

Equations, Modular : for transformation of order 11, XIII, 38—40; for cubic transformation, XIII, 64—5; (*see also* Transformation of Elliptic Functions).

Equations of Motion : in lunar theory, XIII, 206.

Equations, Solutions of : $x^{257}-1=0$, I, 564—6; $\theta^n=1$, and theory of groups, II, 123—30, 131—2, IV, 88—91, X, 610; $x^p-1=0$, XI, 314—6, XII, 72—3; elliptic function solution of $x^3+y^3-1=0$, XII, 35—7; the quaternion $qQ-Qq'=0$, XII, 300—4, 311—3; $(abcd)=(a^2b^2c^2d^2)$, XII, 418—20; $x^{17}-1=0$, XIII, 60—3.

Equations, Theory of : synopsis, X, 3—11; Newton-Fourier method, and imaginary root, XI, 114—21, 143; theorem of Abel's and quintic equation, XI, 132—5; theorem in, XI, 268—9; evolution, XI, 455; in *Ency. Brit.*, XI, 497—521.

Equator: action of gravity at the, IX, 241—3.

Equilibrium: of four forces, V, 540—1, IX, 201; of skew surface, XI, 317—22.

Equimomental Surfaces (*see* Surfaces).

Equipollences: Bellavitis, XII, 473—4.

Equipollent: the term, XII, 473.

Equipotential Curve: III, 258—61.

Essential Singularity of Function: IV, 150.

Eta-Functions: product, XII, 584—6; (*see also* Theta Functions).

Euclid: space of, XI, 434—7; evolution of geometry, XI, 446; proof of I, 47, XI, 557.

Euler, L.: rotation of solid body, I, 237, IV, 566, 567—9, 587, VI, 135—46; involution, I, 259; elliptic functions, I, 366; skew determinants, II, 214; transformation of coordinates, II, 497, IV, 553—7, 587; sums of series, III, 127; indeterminate equations, III, 205—7; polyhedra, IV, 84, 86—7, V, 62—5, 617; *Determinatio Orbitæ Cometæ*, IV, 519, 587; problem of two centres, IV, 525—7, 587; three mutually attracting bodies in right line, IV, 538—9, 587; motion of three bodies, IV, 540, 587; inertia, IV, 562; kinematics of solid body, IV, 580, 587; rotation formulæ, V, 537; differential equation of, VII, 261—2, IX, 592—608, XI, 68—9; binomial theorem, VIII, 463; mathematical tables, IX, 463—6, 471—2, 477—8, 481, 487; theorem on sums of squares, XI, 294; partitions, XI, 360, XII, 219; intersections of cubic curves, XI, 449; gamma function, XI, 535—6; eight-squares theorem, XII, 465; Latin squares, XIII, 55; differential equation of, integrated by Richelot, XIII, 525—9.

Evans, A. B.: Degen's tables, X, 586.

Evectant: of quantics, II, 321.

Evector: of quantics, II, 321.

Evolutes: theory of, V, 473—9; and parallel curves, VIII, 31—45; nodes of, VIII, 329, 351.

Evolution: of geometry, XI, 445—8.

Ewing, J. A.: curve-tracing mechanism, XIII, 505.

Excuboquartic: defined, V, 10, VII, 99; curves, VI, 87—8, XI, 9—13.

Exoscopic: the term, I, 588.

Expansions: in multiple sines and cosines, I, 19—24, 583; in Laplace's coefficients, I, 375—6; of true anomaly, III, 139—42; numerical, IV, 470—2.

Expectation: problem and solution in, X, 587; (*see also* Probability).

Experience and Cognition: XI, 431.

Exponential Functions: and double theta functions, X, 184—5; the term, XI, 524—7.

Extension: in conformal representation, XI, 78.

Extent: the term in seminvariants, XIII, 269, 363.

Extraordinaries: and non-commutative algebras, I, 128—31, 301; the term, XII, 60, 461.

Facients: defined, II, 221, IV, 604, VI, 464.

Factions: the term, IX, 426.

Factorial Expressions: summation of, III, 250—3.

Factorials: developments of, II, 98—101, 594; problems, V, 574, VII, 597; binomial theorem and derivations, VIII, 463—73; maxima of certain functions, VIII, 548—9.

Factors, Special: the term, I, 337.

Facultative: the term, VI, 156, 365; lines of cubic surfaces, VI, 450.

Facultative Points: of Sylvester, XIII, 46.

Family of Quadrics: envelope of, X, 589.

Family of Surfaces: part of orthogonal system, VIII, 269—91.

Fermat, P. de: theorem of, XI, 457, 597, 611, 615—6.

Ferrers, N. M.: conjugate partitions, II, 419; area of conic, III, 143—8; correspondence, X, 290; Legendrian coefficients, XII, 563.

Fiedler, W.: symmetric functions, II, 602.

Figures: for Pascal's theorem, VI, 116—23; mechanical construction of conformable, X, 406; use of arabic, XI, 446.

Finite Differences: formulæ in, III, 132—5, XII, 412—5; electricity on spherical surfaces, IV, 92—8, 99—107; theorem and demonstration, IV, 262—4; general equation of differences of second order, X, 47—9; Stirling's theorem, X, 267—8; table of, XI, 144—7.

Finite Groups (*see* Groups).

Finiteness: of concomitant system of quantic, VII, 334, XI, 272—80.

First Kind: of Abelian integrals, XII, 408—11.

First Order: of differential equations, XII, 432—41; (*see also* Differential Equations).

Five-dimensional Geometry: IX, 79; (*see also* Hypergeometry).

Five-pointic Contact: conic of, IV, 207—39.

Flat: the term in covariants, VIII, 406—8.

Flat-cone: the term, VIII, 102.

Flecnodal Curve: VI, 342; and torse, VI, 345, 582—5.

Flecnode: defined, II, 28—32.

Fleflecnodal Planes: of a surface, X, 262—4.

Fleflecnode: defined, II, 28—32, IX, 264; of curve in transformation of elliptic functions, IX, 170—1.

Flex: the term, V, 521.

Flexure of Skew Surface: XI, 317—22; (*see also* Surfaces).

Floquet, G.: linear differential equations, XII, 394, 402.

Fluctuating Functions: addition to Lord Rayleigh's paper, IX, 19—20.

Fluxions: and Landen, XI, 583.

Focals: of a quadric surface, XIII, 51—4.

Foci: of conics, IV, 505—9, VII, 1—4, 571; theory of, VI, 515—34; and antifoci, problem and solution, VII, 567; locus of, problem and solution, VII, 568.

Focus: the term, VI, 515, IX, 552, XI, 481.

Foot: non-Euclidian geometry, XIII, 483—4.

Forcenex, D. de: multiple algebra, XII, 466.

Forces: equilibrium of, V, 540—1, VII, 91—5; general equation of virtual velocities, IX, 205—8; resultant, X, 589; (*see also* Dynamics).

Forms: cubic, III, 9—12; quadratic, III, 11—12; theory, XI, 604—9.

Formulæ: Jacobi's canonical, for disturbed motion, III, 76—7; in finite differences, III, 132—5; for differentiation, IV, 135—49; distances of point, and tactions, IV, 510—2; signification of elementary one in solid geometry, V, 498—9; integrals for intersections of line and conic, V, 500—4; canonical form of quantics, VI, 153—4; of two sets each of four concyclic points, VI, 509—11, 512—5; focal, and polyzomal curves, VI, 547, 549; of de Jonquières, VII, 41—3; transformation of coordinates, VII, 95—6, 97—8; geodesic lines on ellipsoid, VII, 501—3, 507—10; trigonometric, XII, 108.

Forsyth, A. R.: biographical notice of Cayley, VIII, ix—xliv; addition of elliptic functions, XII, 294.

Foucault, J. B. L.: the earth's rotation, IV, 535, 536, 588.

Fouché, M.: polyhedra, IV, 609.

Fourier, J. B. J.: theorem as to roots of equations, X, 5; theory of equations, XI, 500; (*see also* Newton-Fourier Theorem).

Fourth Dimension: Cayley on, VIII, xxxiii—v.

Fraction-Theorem: Jacobi's, XII, 123—5.

Francais, J. F.: imaginaries, XII, 468.

Franklin, F.: quantics, XIII, 47.

Fresnel, A. J.: wave-surface, IV, 420, XI, 449; wave and tetrahedroid surfaces, X, 252; Sylvester on the optical theory of, XIII, 44.

Frobenius, G. : linear differential equations, XII, 394.

Frost, P. : curves of curvature near umbilicus, VII, 330—1.

Fuchs, L. : Schwarzian derivative and polyhedral functions, XI, 149; linear differential equations, XII, 394, 453.

Function : the term "derivational," I, 63; transformation of bipartite quadric, II, 497—505; relation among derivatives of, X, 590—2; octahedron, XI, 128—9; general theory of, XI, 439—41; linear, XI, 492; in *Ency. Brit.*, XI, 522—42, (introductory, XI, 522—3; known functions, XI, 523—37; functions in general, XI, 537—41; calculus of, XI, 541—2); two invariants of quadri-quadric, XIII, 67—8; on the modular χω, XIII, 338—41.

Functional Determinant : the term, II, 319, IV, 607.

Functional Equation : theorem of Abel, IV, 5—6.

Functions : doubly-periodic, II, 150—63; notation of algebraic, II, 185—8; al(x) of Weierstrass, V, 33—7; homotypical, V, 50; rhizic, IX, 34; tests for dependence of, X, 97; early history of theory, XI, 451—5; values of symmetric, XIII, 318—21; on lacunary, XIII, 415—7; (*see also* Schwarzian Derivative, Generating Functions, Symmetric Functions).

Fundamental Notions : in Mathematics, XI, 434—8, 442—5.

Fuss, P. H. v. : porism formula, II, 90; in-and-circumscribed polygon, II, 140, V, 21—2.

Galileo : and dynamics, XI, 447.

Galois, E. : groups and permutations, II, 134; groups, XI, 133, XIII, 533; theory of numbers, XI, 457, 593, 614; theory of equations, XI, 518—9, 520, 521; biographical notice, XI, 543.

Gamma Function : theory, I, 309—16, 588; a double infinite series, II, 8; the term, XI, 534; (*see also* Definite Integrals).

Gaultier, L. : systems of circles, III, 113; radical axis, XI, 465.

Gauss, J. K. F. : determinants, I, 64; linear transformations, I, 585; attraction of ellipsoids, III, 25—8, 149—53; central forces problem, IV, 520, 588; relative motion, IV, 534; binary quadratic forms, V, 618; *pentagramma mirificum*, VII, 37—8; *Theoria Motus*, VII, 414; geodesic lines on quadric surface, VIII, 156—61; potential of ellipse, IX, 281—2; epispheric integrals, IX, 321, 410—17; mathematical tables, IX, 466—7, 470, 472, 475—7, 488; roots of unity, XI, 60; lemniscate, XI, 64; calculation of log 2, XI, 70; geodesic curvature, XI, 323—4; his theory of surfaces, XI, 331—6; imaginary variables, XI, 439; attractions, XI, 448; theory of equations, XI, 455, 504, 516; theory of numbers, XI, 455, 599, 603; gamma function, XI, 534; biographical notice, XI, 544—5; multiple algebra, XII, 471—2; roots of algebraic equation, XIII, 35; orthomorphosis, XIII, 191; application of surfaces to each other, XIII, 253—64.

Geiser, C. F. : quartic and quintic surfaces, VII, 252; theorem of Cotterill, IX, 506.

Generating Functions : of symmetric functions, Borchardt's, II, 417, 421—3; connected with covariants, IX, 537—42; of quintic, X, 339—400; of quartic, X, 341; of sextic, and binary sextic, X, 394—400; of binary septic, X, 408—21; seminvariants of a given degree, XIII, 306—8.

Generation : of bicircular quartic, X, 223—6.

Generator : the term, V, 169—70, 173—9, 181; nodal, of scrolls, V, 169—70, 179—81.

Generatrix : and the absolute, XIII, 481—9, 501.

Genese, R. W. : theory of envelopes, XI, 50—1.

Genus of Curve : (Geschlecht), after Riemann, V, 467, 517, 619.

Geodesic Curvature : XI, 323—30.

Geodesic Lines : property of, III, 38; on oblate spheroid, VII, 15—25; on ellipsoid, VII, 34—5; in *Ency. Brit.*, XI, 628, 636—7; of pseudosphere, XII, 220—38; wave surfaces, XIII, 252; (*see also* Ellipsoids).

Geodesic Lines, in particular on quadric surface, memoir : VIII, 156—78, 188—99; preliminary formulæ, VIII, 156—8; general theory of them on a surface, VIII, 159—62; circular curves are geodesics, VIII, 162; chief lines not in general geodesics, VIII, 163; special form of geodesic equation, VIII, 163—4;

geodesics on quadric surface, VIII, 164—70; formulæ for position of point, VIII, 170—4; ellipsoid and skew hyperboloid, VIII, 174—8, 188—99; tables, VIII, 196—9.

Geometrical Construction: in optics, X, 28; of heptagon, X, 609.

Geometrical Representation: of elliptic functions, III, 3; of imaginary variables, X, 316—23; of an equation between two variables, XII, 104.

Geometry: of *n* dimensions, I, 55—62; reciprocity, I, 377—82; of quantics, II, 222; of one and two dimensions defined, II, 561—2; of one dimension, II, 563—9, 583—96; of two dimensions, II, 569—83, 586—92; relations of, metrical and descriptive, II, 592; non-Euclidian and hyper-, II, 604—6, VIII, xxxiii—v, 409—13, XII, 220—38; Lobatschewsky's imaginary, V, 471; problem of permutation, V, 493—4; signification of elementary formula, V, 498—9; notion of absolute, V, 550; drawings in, VI, 9; constructive, VII, 26—30; transformation, VII, 121—2; Cayley and Klein on metrical, VIII, xxxvi—vii; hyperbolic, elliptic and parabolic, VIII, xxxvii; Cayley's work in analytical, VIII, xxxviii; formulæ relating to right line, X, 287—9; considerations on solar eclipse, X, 310—5; interpretation of algebraic equations, X, 581; solid, XI, 224; Schubert's numerative, XI, 281—93; Mill on, XI, 432—4; Euclidian, XI, 434—7; Cartesian, XI, 437—9; abstract, XI, 441—2; origin, XI, 445—8; in Greece, XI, 446; evolution of descriptive, XI, 448—9; date of extensions in, XI, 449—51; plane and solid, XI, 450—1; function in, XI, 522—3; interpretation of elliptic function formulæ, XII, 107; d'Alembert Carnot paradox, XII, 305—6; of the compass, XII, 314—7; algebra and logic, XII, 459; (*see also* Hypergeometry: for General Theory, *see* Quantics, sixth memoir).

Geometry, Abstract, Memoir on: VI, 456—69, 596; introductory, VI, 456—7; space, VI, 456—7; general explanations, VI, 457—62, 596; omal relation, order, VI, 463; parametric relations, VI, 463—4; quantics, notations, etc., VI, 464—6; resultant, discriminant, VI, 466—7; consecutive points, tangent omals, VI, 467—9.

Geometry, Analytical, in Ency. Brit.: XI, 546—82; introductory, XI, 546; Part I, pure analytical, XI, 546—67; is descriptive, XI, 552—6; metrical theory, XI, 556—7; equations of right line and circle—transformation of coordinates, XI, 558—61; the conics, XI, 561—4; tangent, normal, circle and radius of curvature, XI, 564—5; coordinates, XI, 566—7; Part II, solid analytical geometry, introductory, XI, 567—9; metrical theory, XI, 570; line, plane, and sphere, XI, 571—2; cylinders, cones, ruled surfaces, XI, 572—3; transformation of coordinates, XI, 573—6; quadric surfaces (paraboloids, ellipsoids, and hyperboloids), XI, 576—9; curves: tangent, osculating plane, curvature, XI, 579—80; surfaces: tangent lines and plane, curvature, XI, 580—2; (*see also* Hypergeometry).

Geometry of Position: theorems in, I, 317—28, 356—61, 414—20, 550—6, 567, 588.

Gergonne, J. D.: caustics, II, 118, 339, 341, 368; polyzomal curves, VI, 520.

Geschlecht: (genus) of curve, after Riemann, V, 467, 517, 619.

Glaisher, J. W. L.: notation for elliptic functions, I, 548; definite integration, VIII, 1; centro-surface of ellipsoid, VIII, 364; report on mathematical tables, IX, 461—99; development of an idea of Eisenstein, X, 58—9; proof of Stirling's theorem, X, 267—8; quadrilateral inscribable in circle, X, 578; log 2, XI, 70; elliptic functions, XI, 73; least factors of numbers, XI, 430; modular function $\chi\omega$, XIII, 338—41; theta and omega functions, XIII, 558—9.

Glide: the term, I, 236.

Glover, J. W.: on theory of groups, XIII, 533.

Goniometry: Cotterill's problem in, X, 295—7.

Göpel, A.: theory of numbers, IV, 41; theta functions, VIII, xlii, X, 464, 499, XII, 363—4; double theta functions and 16-nodal quartic surface, X, 157, 162, 172, 173, 175, 180—1; table of tetrads, 508, 549—51; double theta functions, XI, 454.

Gordan, P.: binary quintic and sextic, VI, 190; irreducible covariants of binary quantic, VII, 334, 341, 348—53; covariants of binary quantic, VIII, 566; finiteness of concomitant systems, X, 286; derivatives, X, 340, 377; Schwarzian derivative and polyhedral functions, XI, 149, 199; finite groups, XI, 237—41; covariantive forms and tables, XI, 272; concomitants of ternary cubic, XI, 343; Abelian functions, XII, 102, 109; icosahedral substitutions, XIII, 552.

Goursat, E.: Kummer's differential equation, XIII, 69—73.

Graduation Curve: the term, VII, 426.

Graham, A.: Pellian equation, XIII, 442.

Grant, R.: report on Catalogue of Memoirs, V, 546—8, 620.

Graphical Construction: in solar eclipses, VII, 390—1, 479—92; geodesic lines on ellipsoid, VII, 507, 510; theory of groups, X, 403—5.

Grassmann, H.: multiple algebra, XII, 465, 480—9.

Gravelius, H.: on non-Euclidian geometry, XIII, 481.

Graves, J. T.: algebraic couples, I, 128; geometry of position, I, 319, 414—20; imaginaries, I, 586.

Gravity: and relative motion, IV, 534—7; action at equator, IX, 241—3; Clairaut's theorem, X, 17—8; effects of theory, XI, 447—8.

Greatheed, S. S.: expansions in multiple sines, I, 583; elliptic motion, IV, 522, 588.

Greek: Cayley's knowledge of, VIII, xxiv; geometry, arithmetic, and algebra, XI, 446.

Green, G.: attractions of ellipsoids, I, 582; attraction of terminated straight line, VII, 32; potentials of polygons and polyhedra, IX, 279; integration of prepotential equation, IX, 320—1, 343, 393—404; attractions, XI, 448.

Greenhill, A. G.: deformation and flexure of surfaces, XI, 66—7.

Greenians: the term, IX, 393.

Greer, H. R.: locus, envelope, and triangle, VI, 72.

Gregory, D. F.: differential and integral calculus, VIII, 272.

Griffiths, J.: series of triangles, VII, 599; curve of sixth order, X, 612; deduction from $y = \sin(A + B + C + ...)$, XII, 58—9.

Ground-curve: and Abel's theorem, XII, 38, 109—216.

Groups: of lines and points, I, 317—28, 356—61, 414—20, 550—6; depending on symbolic equation $\theta^n = 1$, II, 123—30, 131—2, IV, 88—91; theory of, IV, 88—91, X, 324—30, XII, 639—56; the term, IV, 594, 596, VII, 123; rotations of polyhedra, V, 529, 559; Cayley's work at, VIII, xxxiii; theorems on, X, 149—52, 153—4; desiderata and suggestions on theory, X, 401—6; partitions and theory of, XI, 62; homographic transformations, XI, 189—90, 196—208, 237—41; linear transformation of a variable, XI, 237—41; Jacobian sextic, XI, 389, 393—6; the notion, XI, 509—10; Latin squares, XIII, 55—7; substitution groups for two to eight letters, XIII, 117—49; of points, non-existence of a special, XIII, 212; quotient G/H in theory of, XIII, 336—7; illustrations of Sylow's theorems on, XIII, 530—3; of sixty icosahedral substitutions, XIII, 552, 556.

Grunert, J. A.: difference-table, XI, 144.

Gudermann, C.: elliptic integrals, I, 224; transformation of an integral, I, 383; transformation of elliptic functions, III, 2; spherical pendulum, IV, 534, 588; logarithms, VII, 414.

Gudermannian: V, 86—8, 617; Lobatschewsky's imaginary geometry, V, 472; tables of, V, 617.

Gundelfinger, S.: concomitants of ternary cubic, XI, 342—3.

Hall, A.: motion of particle towards attracting centre, IX, 215—7.

Halphen, G. H.: inverse elliptic functions, I, 586; curves in space, V, 613—7; higher singularities of plane curves, V, 619; curves satisfying given conditions, VI, 594—5; classification of curves, XI, 451; invariants of differential equations, XII, 392—3; transformation in elliptic functions, XIII, 29; Sylvester on reciprocants, XIII, 48; reciprocants, XIII, 333, 366, 368—81, 381—98; characteristic n in the theory of curves in space, XIII, 468—72.

Halsted, G. B.: hyperspace and non-Euclidian geometry, II, 606.

Hamilton, Sir W. R.: quaternions, I, 123—6, 238, 335, 586; form of equations of motion, I, 284; homographic transformation of quadrics, II, 105, 133; focal relations, II, 143; problem of three or more bodies, III, 97—103; differential equations, III, 164, IV, 514; method of dynamics, III, 166—74, 200, 202; equations of motion, III, 186; isochronism of circular hodograph, III, 262—5;

essential singularity of function, IV, 105; central forces problem, IV, 520; hodograph, IV, 520; transformation of coordinates, IV, 558—9, 588; ray systems, VIII, 504, XII, 571—5; surface orthogonal to set of lines, IX, 587; system of differential equations, X, 113—8; equations of central orbit, X, 613; on mathematics, XI, 431; algebra and time, XI, 443; conical refraction, XI, 449; multiple algebra, XII, 460, 466, 474—5; Sylvester on Hamiltonian numbers, XIII, 48; (*see also* Differential Equations).

Hammond, J.: theory of tamisage, XI, 409—10; seminvariants, XII, 253; Sylvester's reciprocants, XIII, 47—8, 381, 388; on Hamiltonian numbers, XIII, 48.

Hansen, P. A.: lunar theory, III, 13—24, 291—2; elliptic orbit, III, 95; expansion of true anomaly, III, 140; planetary theory, III, 268—9, IX, 180—3; disturbed elliptic motion, III, 270—1; disturbing function in lunar theory, III, 293, 319—43; variation of plane of planet's orbit, III, 516—8; elliptic motion, IV, 522, 523, 588; relative motion, IV, 536, 588; pendulum, IV, 541, 588; spheroidal trigonometry, IX, 197.

Hargreave, C. J.: on differential equations, VIII, 458.

Harley, R.: equation of differences, IV, 241, 245; symmetric products and quintics, IV, 310—13; quintics, V, 53; a differential equation, VII, 354; theory of equations, XI, 520; invariants, XII, 390—1.

Harmoconic: defined, V, 342.

Harmonic Relations: of two lines or points, II, 96—7; theory of, and two or more quadrics, II, 529—40.

Harmonics: symmetric, II, 555; inscribed, III, 113; reciprocal lines, XIII, 58—9; and non-Euclidian geometry, XIII, 482—9.

Harriot, T.: mathematical discoveries, XI, 437.

Hart, A. S.: cubic curves, IV, 499; relative motion, IV, 535; triple tangent planes, VI, 372, 375; nine-point circle, XIII, 548.

Haughton, S.: inertia, IV, 564—5, 588.

Haupttangenten: (inflexional tangents), VIII, 157.

Heal, W. E.: bitangents of quintic, XIII, 21.

Hearn, G. W.: on a geometrical locus, I, 496; quartic curves, I, 496; quadric curves, V, 262.

Heath, R. S.: non-Euclidian geometry, XIII, 481, 499.

Helmholtz, H. von: hydrodynamical equations, XIII, 6—8.

Hemihedron: the word, X, 328.

Hemipolyhedron: the word, X, 328.

Hensley, P. J.: foci of conics, IV, 505—9.

Heptacron (*see* Polyacra).

Heptagon: construction, X, 609.

Hermite, C.: homographic transformation of quadric into itself, II, 107; elliptic integral and covariants of quartic, II, 191; law of reciprocity, II, 232, 234; skew invariant of quintic, II, 233; transformation of quadric function, II, 499; hyperdeterminants, II, 598—601; elliptic integrals, IV, 68—9; ternary cubics, IV, 326, 330; Tschirnhausen's transformation, IV, 364—7, 375, VI, 165, 170; automorphic transformation, IV, 416; elliptic functions and solution of quintic, IV, 484—9; matrices, V, 438, XII, 367—70, 386; quantics, VI, 147; quintic equation, VI, 170; nodal cubic, VI, 174—6; canonical form of quintic, VI, 177—83; transformation of elliptic functions, VII, 44, IX, 113, XII, 337, 416—7, XIII, 31, 39; reduction of Abelian integrals, X, 214; concomitants of ternary cubic, XI, 342; elliptic functions, XI, 452; theory of equations, XI, 520; Abelian functions, XII, 98; transformation of double theta functions, XII, 358; *H*-product theorem, XII, 584—6; cubic equations, XIII, 349; omega functions, XIII, 558.

Herschel, Sir J. F. W.: finite differences, IV, 95, 107, 262; Brinkley's formulæ, X, 58—9; difference table, XI, 144.

Hesse, L. O.: linear transformations, I, 87, 113—6, 232—3, 584, 585, VI, 22—3, 73; cubic curves, I, 194, II, 399; involution, I, 259; hyperdeterminants, I, 354; abstract of memoir on quadric surfaces, I, 425—7; inflexions of cubic, I, 584, II, 29, III, 48; on elimination, IV, 3; on inflexions, IV, 186, XI, 473; double tangents, IV, 186—7, 343—6; cubic forms, IV, 353; geometric transformation, VII, 121—2; bitangents of quartic curve, VII, 123—4; double theta functions, X, 177; triple theta functions, X, 446, 448, 451; the thirty-four concomitants of ternary cubic, XI, 342; bitangents of a plane quartic, XII, 76.

Hessians: and Eisenstein's theorem, I, 585; defined, II, 319, IV, 607, XI, 471, 474; and Pippian, II, 381—2, 383—95, 416; the sign for, II, 541; the quadricovariant, II, 545; of quaternary functions, IX, 90—3; of a quartic surface, X, 274—7; (*see also* Quantics).

Hexagon: and conic, problem, V, 576; theorem of inscribed, XI, 556.

Hexagram (*see* Pascal's theorem).

Hexahedron: edges of, problem, X, 613; automorphic function for, XI, 184.

Hierholzer, C.: cones satisfying six conditions, VIII, 99; surface of eighth order, VIII, 401.

Higher Singularities of Curves (*see* Singularities).

Hilbert, D.: curves in space, V, 614; abstract geometry, VI, 596.

Hills: altitude, and roots of algebraic equation, XIII, 33—7.

Hirsch, Meyer: algebra of, II, 417, 440; partitions, VII, 577.

Hirst, T. A.: negative pedals, IV, 164.

History: importance of mathematical, VIII, xii.

Hodograph: isochronism of circular, III, 262—5; Hamilton's, IV, 520; and pedal curves, V, 113.

Hölder, O.: theory of groups, XIII, 336, 533.

Holditch, H.: caustics, II, 363.

Holomorphic: the term, XI, 81.

Homaloids: prepotentials of, IX, 408—9.

Homographic Figures: matrices, II, 219; theorem, IV, 442—5.

Homographic Function: distribution of electricity, XI, 2—6; and matrix of the second order, XI, 252—7.

Homographic Relations: and theory of numbers, IX, 613; function of, X, 298—306; powers of, X, 305—6, 307—9.

Homographic Transformation: of quadric surfaces, II, 105—112; of single theta functions, XII, 337—43; (*see also* Groups, Transformations).

Homographies: correspondence with rotations, X, 153—4, XI, 237—8.

Homography: different classes of, II, 219—220; and quadrics, II, 536—40; ternary quantics and theory of, II, 565—9, 578; of ranges and pencils, II, 578.

Homology: of conics, I, 523, 557; the term, I, 557—8; of sets, III, 35.

Homotypic: the term, VII, 123.

Homotypical: the term, V, 50.

Hopkins, W.: attractions and multiple integrals, I, 582.

Horner, W. G.: theory of equations, XI, 499, 502.

Hoüel, J.: equations of motion, III, 169, 170.

Hudson, C.: equal roots of equations, XI, 405—7.

Hydrodynamics: note on equations of, XIII, 6—8.

Hydrogen: trees of, IX, 427—60, 544—5.

Hyperbola: rectangular, III, 254, V, 554; classification, V, 354—400, VIII, xxxviii; arc of, XI, 337—9; and cubic curves, XI, 478; in *Ency. Brit.*, XI, 561—4; focals of quadric surface, XIII, 54; and epitrochoid, XIII, 82—7; tactions, XIII, 150—69; (*see also* Conics).

Hyperboloid: coordinates of, V, 72; and scrolls, VII, 65; and tractor, VII, 73—5; passing through three given lines, VII, 86—8; Mercator's projection, VIII, 568, IX, 237—40; deformation, XI, 66—7; in *Ency. Brit.*, XI, 576—9; confocal Cartesians and right lines of, XII, 587—9.

Hyperdeterminants: the term, I, 81, 95, 114, 585; note on, I, 352—5, 588; a system of certain formulæ, I, 533; theory, I, 577—9; theory of permutants, II, 19; theory of intermutants, II, 26; quantics, II, 225; theory of seminvariants, XII, 344; Sylvester's work in, XIII, 46; an identity, XIII, 210—11; (*see also* Covariants, Invariants).

Hyperdimensional Space: quadrics in, IX, 79—83; (*see also* Hypergeometry, Hyperspace).

Hyperelliptic Functions: trisection of, VI, 594; and theta functions, X, 162—5, 166—79, 184—214, 551—5; and triple theta functions, X, 432—6; addition-theorem, X, 455—62; the term, XI, 533—4; and nodal quartics, XII, 196—208; (*see also* Theta-Functions).

Hyperelliptic Integrals: of first order, XII, 98—9.

Hypergeometric Series: summation of a certain factorial expression, III, 250—3; theorem, III, 268—9; differential equations, XI, 17—25; note on, XI, 125—7; and Schwarzian derivative, XI, 176—9.

Hypergeometry: of n dimensions, I, 55—62; a branch of mathematics, VIII, xxxiii—v; five-dimensional, IX, 79—83; and quadric surfaces, IX, 246—9; 21 coordinates of conic in space, XI, 82—3; Sylvester's work in, XIII, 46; (*see also* Hyperspace, Prepotentials).

Hyperspace: and quantics, II, 222; and non-Euclidian geometry, II, 606; representation by means of, VI, 198; of four dimensions, special theorem, IX, 246—9; (*see also* Hypergeometry).

Icosahedra: construction, IV, 81—2; axial systems, V, 531—9; Klein on rotations of, X, 153; as regular solids, X, 270—3; automorphic function, XI, 169, 179—83, 185, 212—6.

Icosahedral Substitutions (*see* Substitutions).

Ideal: the term, VI, 483.

Ideal Numbers: XI, 456.

Idem: defined, XII, 66.

Idempotent: the term, XII, 61.

Identities: cubic, V, 597; trigonometrical, VIII, 525, XI, 38, XIII, 538—40; elliptic transcendent, VIII, 564; a transcendental, XI, 37; algebraic, XI, 63—4, 130—1, XIII, 76—8; a hyperdeterminant, XIII, 210—11.

Imaginaries: on an octuple system of, I, 301; eight-square, XI, 368—71, XII, 465; the term, XI, 439; theory of equations, XI, 502—6; and function, XI, 523; associative, XII, 61, 105—6; perpendicularity, XII, 466—72; roots of equation, XIII, 36; Sylvester's work at, XIII, 46; quaternions, XIII, 542.

Imaginary Quantities: logarithms, VI, 14—8; geometrical construction relating to, XI, 258—60.

Immit: defined, IV, 109.

Improper: conditions for curves, VI, 193.

Increment: the term, VI, 468.

Indefinite: applied to integration, IX, 500—3; the term, XIII, 290.

Indeterminate Equations: problem in indeterminate analysis, III, 205—7.

Index: to philosophic memoirs, report on, V, 546—8, 620.

Indicial Equation: of differential equation, XII, 398, 453.

Indicial Function: of differential equation, XII, 398, 401.

Inertia: axes and moments of, IV, 478—80, 559—66.

Ineunt: defined, II, 574, V, 521, VI, 469; non-Euclidian geometry, XIII, 489.

Infinitesimal Rotations: VI, 24—6.

Infinity: in geometry, XI, 464.

Inflexional Tangents: and geodesic lines, VIII, 157; (*see also* Tangents).

Inflexions: of cubical divergent parabolas, V, 284—8; of cubic curve, I, 584, III, 48; Hesse on, IV, 186, V, 493—4, XI, 473; of curves, XI, 471—3, 480.

Integral Calculus: some formulæ of, I, 309—16, 588; transformation, I, 383; Picard's memoir on, XII, 408—11.

Integral Functions: Legendre's coefficients, I, 375—6; the term, IV, 603—4, XI, 523; prepotential surface, IX, 321—30, 330—4, 352—9; potential solid, IX, 334—7; epispheric, IX, 410—17; reduction of transcendental, x, 214—22; hyperelliptic, of first order, xii, 98—9; regular, of differential equation, xii, 395—6; number of, xii, 399; subregular of differential equation, xii, 444—52; (*see also* Abelian Integrals, Definite Integrals, Elliptic Integrals, Transformations).

Integrals: transformation of double, ix, 250—2; of differential equations of first order, x, 19.

Integration: a supposed new, vii, 36; theorem of, vii, 588; by series of differential equations, viii, 458—62; a process of, ix, 257—8, x, 15, 29; indefinite, ix, 500—3; Aronhold's formula, x, 12—14; of Euler's equation, xi, 68—9.

Integrator: mechanical, xi, 52—4.

Intercalation: root-limitation, ix, 22—7; for right line, ix, 28—33; Sylvester's theory of, xiii, 46.

Intermediates: of two quantics, defined, ii, 515; of binary quartic, ii, 549; and ternary cubics, iv, 326.

Intermutants: the term, ii, 19, 26, iv, 594, 600.

Interpolation: Smith's Prize dissertation, viii, 551—5.

Intersect-developable: of two quadrics, i, 486—95.

Intersections: the term, vii, 546; of two curves, ix, 21, xii, 117—20; of cubic and line, xii, 100.

Invariable Plane: and rotation of solid body, i, 237, vi, 142.

Invariants: the term, i, 577, 589, ii, 176, 224, iv, 594, 605, xiii, 46; and discriminants, i, 584; determined by differential equations, ii, 164—78; and roots, ii, 176; differential equation satisfied by, ii, 176—8; and binary quantics, ii, 266—8; of quartic, and covariants of cubic, analogous, ii, 553; bibliography, ii, 598—601; of biternary quantics, iv, 349—58; 18-thic of quintic in terms of roots, vi, 154—6; Cayley founder of, viii, xxviii—xxx; his work, viii, xxx—xxxii; and transformation of quantics, viii, 385—7; quadratic transformation of a binary form, viii, 398—400; identical equation connected with theory, ix, 52—5; Hessian of quaternary function, ix, 90—3; minimum N. G. F. of binary septic, x, 408—21; stereographic projection, xi, 187—9; in geometry, xi, 474; of a linear differential equation, xii, 390—3; Sylvester's work at, xiii, 46, 47; two, of quadri-quadric function, xiii, 67—8; differential, and reciprocants, xiii, 366—404; Pfaff-, xiii, 405—14; (*see also* Covariants, Linear Transformations, Seminvariants).

Invariants and Covariants: xii, 22—9; standard solutions of system of linear equations, xii, 19—21; finite number of the covariants of a binary quantic, xii, 558.

Inversion: of quadric surface, viii, 67—71; note on, ix, 18.

Inverts: quadric function of, xi, 153—6.

Involutant: of two binary matrices, xiii, 74—5.

Involution: theory of geometrical, i, 259—66, 587; and two or more quadrics, ii, 529—40; of six lines, iv, 582, vii, 66, 85, 95; lines in, v, 1—3; theory, v, 295—313; the term, vi, 460; of four circles, vi, 505—8; and ternary quadrics, xiii, 350—3.

Involution of Cubic Curves, Memoir: v, 313—53, vii, 238; explanations, definitions, and results, v, 314—8; general formulæ for critic centres, v, 318—9; twofold and one-with-twofold centre, v, 319—24; tangents at a node, v, 325—8; triangle of critic centres, v, 328—9; the three-centre conic, v, 329—36, 337—8; transformation equation of cubic, v, 339—41; cubic locus, harmoconics, and harmonic conic, v, 341—5; miscellaneous, v, 345—53.

Irrational: and subrational, ix, 315.

Irreducible: the term, vii, 336, xii, 23.

Irreducible Concomitants: of quintic, x, 342.

Irreducible Covariants: and invariants, ii, 250.

Irreducible Syzygies (*see* Syzygies).

Irregular: the term, vi, 457, 459.

Iseccentric Lines: and planet's orbit, vii, 468.

Isobaric: the term, xiii, 266.

Isobarism: of covariants, ii, 233.

Isochronic: the term, nodal and cuspidal, vii, 473.

Isochronism: of circular hodograph, III, 262—5.

Isomers: Mathematical theory of, IX, 202—4.

Isoparametric Lines: and planet's orbit, VII, 467.

Isoperimetrical Problem: VII, 263.

Isothermals: of Meyer, XIII, 175.

Jacobi, K. G. J.: theory of algebraic curves, I, 53; determinants, I, 63, 64, 66; quaternions, I, 126, 127, 586; inverse elliptic functions, I, 132, 136, 152, 156, 162, 180; rotation of solid body, I, 238, IV, 575, 576—7, 579; involution, I, 259, 263; definite integral, I, 270—1; dynamical differential equations, I, 276—9; elliptic functions, I, 290—300, 507, 587, XI, 452; demonstration of theorem on focal lines, I, 362—3; differential equations of Abelian functions, I, 366—9; skew determinants, I, 411; simultaneous linear transformations, I, 428; transformation of integrals, I, 440, 442; attraction of ellipsoids, I, 511—8; solution of equation $x^{257} - 1 = 0$, I, 564; Lagrange's theorem, II, 7; geometrical representation of elliptic integral, II, 56; in-and-circumscribed polygon, II, 141; partition of numbers, II, 248; lunar theory, III, 13; canonical formulæ for disturbed motion, III, 76—7; problem of three or more bodies, III, 102; transformation, III, 129; finite differences, III, 132—5; complete integral of partial differential equation, III, 166; planetary theory, III, 173; calculus of variations and differential equations, III, 174—84, 200, 202; *De Motu Puncti singularis*, III, 182—3, 202; problem of three bodies, III, 519—21, IV, 541, 548—51, 589, V, 23, VI, 183; *Theoria Novi Multiplicatoris*, III, 183—5; theory of ideal coordinates, III, 185; Encke's memoir *über die speciellen Störungen*, III, 179—80; in-and-circumscribed triangle, III, 236; disturbed elliptic motion, III, 270—1; canonical system of formulæ, III, 290; reversion of series, IV, 30—7; transformation of elliptic integrals, IV, 60, 64; double tangents, IV, 187; conics, IV, 207; Pfaff's problem, IV, 359—63; central forces problem, IV, 520, 589; *Nova Methodus*, IV, 515, 521, 589; elliptic motion, IV, 522, 589; problem of two centres, IV, 530, 532, 589; motion of a single particle, IV, 537—8, 589; motion in resisting medium, IV, 541, 589; motion of a point, IV, 547, 589; elimination of nodes, IV, 551, 589; transformation of coordinates, IV, 554, 557, 589; Weierstrass's function al (x), V, 34—5; transformation of elliptic functions, V, 472, IX, 103, 113—75, XII, 59, 505—34; *Canon Arithmeticus*, VI, 83—6, XI, 85, 86; the Jacobian relation, VI, 467; geodesic lines on ellipsoid, VII, 493; transformation of double integral, IX, 250—2, 254; epispheric integrals, IX, 321, 410—17; mathematical tables, IX, 472—3, 484—5; series, X, 25—7; Poisson's theorem, X, 108—9, 110—3; theta functions, X, 156, 473, 478, 490, 496—7, XI, 41—6, XIII, 559; roots of unity, XI, 58—60; Schwarzian derivative and polyhedral functions, XI, 149; hypergeometric series, XI, 178; Landen's theorem, XI, 339; Abelian functions, XI, 454; theory of numbers, XI, 602; theorem in simultaneous equations, XII, 39; fraction theorem, XII, 123—5; Weierstrassian and Jacobian elliptic functions, XII, 425—7; sextic equation, XI, 389—401, XII, 493—9; sums of two series, XIII, 50; modular equations, XIII, 64; characteristic n and curves in space, XIII, 469; sextic resolvent equations, XIII, 473—9.

Jacobian: defined, II, 319, IV, 607; of two quantics, II, 517; theory of, and polyzomal curves, VI, 566—8; the extended notion, VII, 134; of surfaces, VII, 134—6; of two curves, IX, 21; of quadric surfaces, X, 568; of six points, X, 613; rational transformation, XIII, 116; (*see also* Quantics).

Jefferson, T.: founder of Virginian mathematical professorship, XIII, 43.

Jeffery, H. M.: on quartic curves, XI, 408.

Jellett, J. H.: theorem of, on attractions, I, 388—91.

Jenkins, M.: Arbogast's method of derivations, IV, 609; problem in factorials, VII, 597.

Jerrard, G. B.: quintics, IV, 391—4, V, 50—4, 77, 89; theory of equations, XI, 520.

Joachimsthal, F.: theory of covariants, II, 177, 234; theorem of, II, 521, XII, 594—5, 601, 629; normals of a conic, IV, 74—7; attraction of terminated straight line, VII, 33.

Johnson, W. W.: matrices, XI, 252.

Jordan, C.: trees, X, 599, XI, 366; substitutions, XI, 455; theory of equations, XI, 520—1; uniform convergence, XIII, 343—4.

Joubert, P.: transformation of elliptic functions, IX, 113, 138.

Kant, I.: cognition and space, XI, 431; negative magnitude, XI, 434.

Kelland, P.: quaternions, XII, 475.

Kelvin, Lord: equimomental surfaces, I, 253—4; definite integral, I, 270; differentiation and evaluation of definite integrals, I, 587; relative motion, IV, 535, 593; inertia, IV, 565, 566, 593; attraction of terminated straight line, VII, 33; report on mathematical tables, IX, 461—99; distribution of electricity, XI, 6.

Kempe, A. B.: theory of groups, XII, 639.

Kenogram: the term, IX, 202—3.

Kepler, J.: problem of, IV, 521; and ellipse, XI, 447.

Key, T. H.: Professor of Mathematics, University of Virginia, XIII, 43.

Kiepert, L.: transformation of elliptic functions, XII, 490—2.

Kinematics: of solid body, IV, 580—2; six coordinates of a line, VII, 89—95; of a plane, XI, 103—110, XIII, 505—16.

Kirkman, T. P.: schoolgirl problem, I, 483, 589; geometry of position, I, 550—6; sums of squares, II, 49—52; double summation of factorial expression, III, 250—3; autopolar polyhedra, IV, 85, 182; enumeration of polyhedra, V, 38; partitions of a polygon, XIII, 93.

Klein, F.: theory of distance, II, 604, VIII, xxxvi—vii; coordinates in non-Euclidian geometry, II, 604—6; non-Euclidian geometry, VIII, 409—13, XIII, 481; correspondence of homographies and rotations, X, 153; a system of quadric surfaces, X, 269; octahedral function, XI, 128; Schwarzian derivative and polyhedral functions, XI, 149, 151, 179, 183, 185; his classes of geometry, XI, 436; transformation of elliptic functions, XII, 29, 46, 490, 547.

Knots: of trees, III, 243, IX, 429—60, XI, 365—7, XIII, 26—8; in topography, defined, IV, 104.

Königsberger, L.: transformation of elliptic functions, IX, 113, 140; theta functions, X, 499.

Korndörfer, G.: quartic and quintic surfaces, VII, 246, 249, 252.

Kowalski, M.: orbit of Neptune, IX, 180.

Kreistheilung: (cyclotomy), XI, 58, 86.

Kronecker, L.: equation of differences, IV, 609—16; quintics, V, 55; theory of equations, XI, 520; sextic resolvent equations, XIII, 473—9.

Kummer, E. E.: solution of equation $x^{257} - 1 = 0$, I, 564; wave surfaces, I, 587; composition of numbers, IV, 70—1, 78—9; Steiner's quartic surface, V, 423; 16-nodal quartic surfaces, V, 431—7, VII, 126—7, X, 437; quartic surfaces, VII, 134, 176, 313, IX, 71; quintic surfaces, VII, 245, 247, 252; nodal quartic surfaces, VII, 264—297; surface of, VIII, 69; prepotentials, IX, 364; mathematical tables, IX, 494; double theta functions and 16-nodal quartic surface, X, 158, 161; octic surface, X, 252; theta functions, X, 499; table of hexads, X, 506, 538, 552; hypergeometric series, XI, 17—25, 177; Schwarzian derivative and polyhedral functions, XI, 149; theory of numbers, XI, 456; systems of rays, XII, 573; differential equation of third order, XIII, 69—73.

Lachlan, R.: circles and spheres, I, 581, XIII, 13.

Lacroix, S. F.: transformation of coordinates, IV, 557; imaginaries, XII, 468.

Lacunary Functions: XIII, 415—7.

Ladd, Christine: the Pascal hexagram, VI, 594.

Lagrange, J. L.: theorem of expansion, I, 40—2, 584, II, 1—7, III, 141; determinants, I, 64; variation of parameters, I, 243; dynamical differential equations, I, 279; elliptic functions, I, 366; distances of points, I, 581; attractions and multiple integrals, I, 581—2; caustics, II, 353; problem of two fixed centres, III, 104—10, IV, 527, 528—9, 589; sums of series, III, 124; *Mécanique Analytique*, III,

14—2

157—8, 201, 202; equations of motion, III, 158, 200, IX, 198—200; planetary theory, III, 159—61, 162—3, 201; variation of arbitrary constants in mechanical problems, III, 161—5, 200, 201; coefficient (a, b) of, III, 163; Hamilton's method of dynamics, III, 171—3, 200; disturbed elliptic motion, III, 271—2; equations of differences, IV, 240, 252; resolvents, IV, 309; central forces problem, IV, 519—20, 589; elliptic motion, IV, 521—2, 589; expansion of anomalies, IV, 522; spherical pendulum, IV, 532—3, 589; rotation of solid body, IV, 566, 569, 589; homotypical functions, V, 50; invariable plane, VI, 142; invariants, VIII, xxx; demonstration of Taylor's theorem, VIII, 493—5, 519; virtual velocities, IX, 205—8; prime roots of unity, XI, 57; Schwarzian derivative, XI, 149; theory of equations, XI, 455, 498—500, 514—5, 517, 520; envelopes, XI, 475; minimal surfaces, XI, 638; five points in space, XII, 581—3; theorem of expansion and partitions of polygon, XIII, 113; Waring's formula, XIII, 215—6; reciprocants, XIII, 366; Richelot's integral of Euler's equation, XIII, 526.

Laguerre, E.: theory of matrices, II, 604.

Lamb, H.: conformal representation, X, 290.

Lambdaic: defined, II, 523, IV, 49, 53; of binary quartic, II, 550.

Lambert, J. H.: theorem on circular hodograph of, III, 262—5; theorem for elliptic motion, III, 562—5, VII, 387—9; central forces problem, IV, 519, 520, 589; equation of planet's orbit from three observations, VII, 412—5.

Lamé, G.: curvilinear coordinates, VIII, 146, XI, 637, XII, 17; orthogonal surfaces, VIII, 280, 291, 292.

Lamp, Milner's: a differential equation, and construction of, XIII, 3—5.

Lancret, M. A.: curves of curvature, XII, 601.

Landen, J.: theorem of, in elliptic functions, XI, 337—9; biographical notice, XI, 583—4.

Languages: Cayley's knowledge of, VIII, xxiii.

Laplace, P. S.: on Lagrange's theorem, I, 42, II, 7; determinants, I, 63; functions of, I, 397—401, 588; attraction of ellipsoids, I, 581, III, 53—65, 567; planetary theory, III, 159, 201; disturbed elliptic motion, III, 505, 510—11; on secular variation, III, 568; elliptic motion, IV, 524, 589; relative motion, IV, 534, 536, 589; motion of three bodies, IV, 540—1, 589; prepotentials, IX, 393; finite differences, XII, 412.

Last Multiplier: IV, 530, 547, 551, 590.

Latin Squares: XIII, 55—7.

Latitude: parametric, VII, 16, IX, 238.

Lattice: in theory of numbers, III, 40.

Laverty, W. H.: systems of equations, VII, 578.

Law, The: Cayley's work at, VIII, xiii—xv, xix.

Lebesgue, V. A.: determinants, I, 63.

Lectures: delivered by Cayley, VIII, xvi—xvii, xlv.

Lefort, F.: elliptic motion, IV, 522, 589.

Left-handed: circuits in root-limitation, IX, 22—3.

Legendre, A. M.: elliptic functions, I, 136, 156, 507, V, 618, XI, 452, 537, 584, XII, 35—7; elliptic integrals, I, 224; coefficients of, I, 375—6; attraction of ellipsoid, I, 432—7, 442; functions of, IV, 99, 106; rectilinear motion, IV, 516, 590; central forces problem, IV, 521, 590; problem of two centres, IV, 530, 590; rotation of solid body, IV, 570, 590; geodesic lines on oblate spheroid, VII, 15—25; reduced latitude, VII, 16; orbit of planet from three observations, VII, 414; mathematical tables, IX, 467—8, 478, 487; Landen's theorem, XI, 339; theory of numbers, XI, 455, 602—4, 616, XII, 35—7; second kind of coefficients, XII, 562—3; gamma function, XIII, 49.

Lejeune-Dirichlet, P. G.: multiple integrals, I, 195, 582—3; integration, I, 588; theorem of, II, 10, 47—8; binary quadratic forms, V, 141; prepotentials, IX, 321, 417—23; attractions, XI, 448; theory of numbers, XI, 456, 616.

Lemniscate Function: XI, 65; and orthomorphosis, XIII, 191—205.

Letters: substitution groups for two to eight, XIII, 117—49.

Leverrier, U. J. J.: disturbing function in planetary theory, III, 321, VII, 511—27; elliptic motion, III, 361, 362, IV, 523, 590; position of orbit in planetary theory, VII, 545.

Lévy, M.: orthogonal surfaces, VIII, 269, 569—70; Dupin's theorem, IX, 85.

Light, Polarized: MacCullagh's theorem, IV, 12—20.

Limaçon of Pascal: I, 480, XI, 477.

Lindemann, F.: non-Euclidian geometry, XIII, 481.

Linear: and omal relations in abstract geometry, VI, 463.

Linear Differential Equations: invariants of, XII, 390—3; general theory, XII, 394—402, 444—52; decomposition, XII, 403—7.

Linear Equations: and determinants, XI, 490; standard solutions of system of, XII, 19—21.

Linear Function: the term, XI, 492.

Linear Quantics (*see* Quantics).

Linear Substitutions: note on a function in, X, 307—9.

Linear Transformations: theory of, I, 80—94, 95—112, 117, 584, 585; Eisenstein's and Hesse's formulæ, I, 113—6, 585; homogeneous functions of third order with three variables, I, 230—3; hyperdeterminants, I, 352—5, 577—9, 588, 589; theory of permutations, I, 423—4; simultaneous, of two homogeneous functions of second order, I, 428—31; theory of permutants, II, 19—23; the term, IV, 594, 605; of elliptic integrals, IX, 618—21; of theta functions, XII, 50—5; Sylvester's work in, XIII, 46; (*see also* Covariants, Invariants, Quantics).

Line-geometry: and congruences, XIII, 228—30; (*see also* Coordinates, Lines).

Lineo-linear: the term, II, 517, IV, 604, VI, 464.

Lineo-linear Transformation: between planes, VII, 215—6, 236—8.

Line-pairs: the term, VI, 206, 209, 210, 211; through three given points and touching given conic, VI, 201, 244, 594.

Line-pair-point: the term, VI, 202, 210, 211, 269, 594—5.

Lines: on cubic surfaces, I, 445—56, VIII, 371—6; harmonic relation of two, II, 96—7; of cubic curve, II, 382; satellite, II, 383, V, 359; formulæ, II, 405—9; line, plane and point, defined, II, 561—2; contour and slope, IV, 108—11, 609; cubic centres and cones, IV, 173—8, 179—81; geometry of, IV, 446—55, 616—8; involution, IV, 582, V, 1—3; cubic centres of three lines and a line, V, 73—6; theorem of conic and triangle, V, 100—2; intersections of pencils of four and two, V, 484—6; formulæ for intersections of line and conic, V, 500—4; circle and parabola, problem, V, 607; notation in Pascal's theorem, VI, 116—23; facultative, VI, 365—6, 450; dot-notation for, and planes and cubic surfaces, VI, 365—6, 373—449; twenty-seven on cubic surface, VI, 371—87; attraction of terminated straight, VII, 31—3; five on cubic surface, VII, 177—8; homographic transformation, VII, 193—7; spherogram and isoparametric, VII, 467—8; iseccentric and *e*-spherogram, VII, 468—70; isochronic and time spherogram, VII, 470—7; Cayley's work on six coordinates of, VIII, XXXV; potentials of, IX, 278—80; formulæ relating to right, X, 287—9; and points, X, 570; and conics, X, 602; contact with a surface, XI, 281—93; Mill on, XI, 432—3; non-Euclidian geometry, XI, 437, XIII, 480—504; evolution of theory of curves, XI, 450—1; singularities of curves, XI, 468; in *Ency. Brit.*, XI, 548, 571—2; equation of right, XI, 558—61; and surface, XI, 629; Mascheroni's geometry of the compass, XII, 314—7; reciprocal, XIII, 58—9, 481; identity relating to six coordinates of a line, XIII, 76—8; and notion of plane curve of given order, XIII, 79—80; syzygetic relations, XIII, 224—7; of striction, on skew surface, XIII, 232—7; system of in a plane, and their orthotomic circle, XIII, 346—7; and point, distance, XIII, 495—7; theory of two lines, XIII, 497—504; (*see also* Coordinates, Curvature, Geodesic Lines).

Line Systems: two-dimensional geometry, II, 569—83.

Link: the term, V, 521, VII, 183, XIII, 506.

Linkage: the MacMahon, XIII, 265, 292, 293, 298—301.

Link-work: X, 407.

Liouville, J.: integration of differential equations, III, 180, 191—2; equations of motion, III, 185; Bour's memoir, III, 199, 204; a definite integral, IV, 28—9; elliptic motion, IV, 522, 590; problem of two centres, IV, 530; motion of three bodies, IV, 541, 590; point moving in plane, IV, 542—6, 590; Abelian integrals, IV, 546, 590; on roots of equations, IX, 39; indefinite integration, IX, 500—3; geodesic curvature, XI, 323, 328.

Lipschitz, R.: on Lagrange's equations, IX, 110—2.

Listing, J. B.: partitions of a close, V, 617, VI, 22; theorem of, VIII, 540—7.

Lobatschewsky, N. I.: geometry of, V, 471—2, VIII, xxxvii, XI, 436, XII, 220—38.

Locus: defined, IV, 458, VI, 458, 460; from two conics, VI, 27—34; in relation to triangle, VI, 53—64; and envelope in regard to triangle, VI, 72—82; *in solido*, problem and solution, VII, 599; *in plano*, problem and solution, VII, 599; the terms nodal-, cuspidal-, envelope-, tac-, VIII, 533—4; in *Ency. Brit.*, XI, 585.

Logarithms: theory of, III, 208—13, 222—8; of imaginary quantities, VI, 14—8; Pineto's tables, VIII, 95—6; calculation of log 2, XI, 70; origin, XI, 447; function in, XI, 526—7.

Logic: of characteristics, III, 51—2; calculus of, VIII, 65—6; Mill on, XI, 432—4; recent developments, XI, 458—9; geometry and algebra, XII, 459.

Logistic: and algebra, V, 293—4.

London Mathematical Society: Sylvester's connexion with, XIII, 45.

Lottner, C. L. E.: motion of solid body, IV, 583, 590.

Lubbock, Sir J. W.: theory of the Moon, III, 309—10.

Lunar Theory: Hansen's, III, 13—24; disturbed elliptic motion, III, 270—92; development of disturbing function, III, 293—318, 319—43, VII, 511—27; tables, III, 299—308; theorem of Jacobi, III, 519—21; Plana's, VII, 357—60, 361—6, 367—70, 371—4, 375—6; Delaunay's l, g, h, and $h+g$, VII, 528—33, 534; pair of differential equations in, VII, 535—6, 537—40; Newcomb's work, IX, 179—80; note on, XIII, 206—9; (*see also* Elliptic Motion, Disturbed).

Lüroth, J.: six coordinates of a line, VII, 66.

Luther, E.: theory of equations, XI, 520.

MacAulay, A.: quaternions and hydrodynamical equations, XIII, 8.

MacClintock, E.: quintics, IV, 609—16.

MacCullagh, J.: polar plane theorem, IV, 12—20; spherical conics, IV, 428; inertia, IV, 564—5, 588; rotation of solid body, IV, 574, 588.

MacDowell, J.: triangle and circle, V, 564.

Maclaurin, C.: cubic curves, I, 586; attraction of ellipsoids, III, 27, 62, 155; pedal curves, V, 113—4.

MacMahon, P. A.: symmetric functions, II, 603; seminvariants and symmetric functions, VIII, xxxi, XIII, 286—8, 290, 292, 293, 302; a differential equation, XII, 30—2; seminvariants, XII, 239, 254, 261, 275, 349, XIII, 265; a differential operator, XII, 318; Waring's formula for form of equation adopted by, XIII, 214; reciprocants and differential invariants, XIII, 399—404.

Magic Squares: X, 38.

Magnetism: Sabine's work, XI, 430; Gauss's work, XI, 544.

Maillard, M. S.: penultimate forms of curves, VIII, 258; degenerate forms of curves, XI, 220; systems of curves, XI, 487.

Major Function in Abel's theorem: XII, 111, 132—56.

Malet, J. C.: invariants, XII, 390—1.

Malfatti, G. F.: sextic resolvent equations, XIII, 473.

Malfatti's Problem: and system of equations, I, 465—70; Steiner's extension, II, 57—86, 593; Schellbach's solution, III, 44—7; system of equations connected with, IX, 546—50.

Malus, É. L.: systems of rays, XII, 573.

Mannheim, A.: epitrochoid, XIII, 81.

Manuscript of·Cayley: facsimile of, on frontispiece, VIII.

Maps: surface representation on plane, VIII, 538—9; colouring, XI, 7—8; projections, XI, 448.

Mars: Newcomb on observations of, IX, 177.

Marsano, G. B.: mathematical tables, IX, 481—3.

Martin, A.: integration, X, 15, 29; probabilities, X, 600; Pellian equation, XIII, 467.

Mascheroni, L.: geometry of the compass, XII, 314—7.

Masères, F.: algebraic equations, IV, 171.

Mathematical Society of London: Sylvester's connexion with, XIII, 45.

Mathematical Tables (*see* Tables).

Mathematics: recent terminology, IV, 594—608; Mill on, XI, 432—4; relation to physics, XI, 444, 449; extent, XI, 449; Sylvester on its relation to music, XIII, 45.

Matrices: notation, II, 185—8, XI, 243; constituents being linear functions of one variable, II, 216—20; unity, II, 477; theory of, II, 475—96, 604, V, 438—48; which satisfy algebraic equation of own order, II, 483; convertible, II, 488; rectangular, II, 494—6; in automorphic linear transformation, II, 497—505; a linear system, II, 604; for reduction of quintic to Jerrard's form, IV, 392; the term, IV, 594, 601—2; Bezoutic, IV, 607; Cayley's method of verification, VIII, xxvii; Cayley the founder of theory, VIII, xxxii—iii; transformation of coordinates, XI, 136—42; of order two and the homographic function, XI, 252—7; quaternions, XII, 303, 311, 479; the equation, $qQ - Qq' = 0$, XII, 311—3; and theta function transformation, XII, 367—72, 386—9; quintic, XII, 376—80; Sylvester's theory of the corpus, XIII, 47; involutant of two binary, XIII, 74—5; six coordinates of a line, XIII, 76—8; note on a theorem in, XIII, 114; and sixty icosahedral substitutions, XIII, 552—7.

Maurice, F.: variation of arbitrary constants, III, 166, 202.

Maxima: of certain factorial functions, VIII, 548—9.

Maxima and Minima: of functions of three variables, I, 228—9; theorem in, IX, 40—1; theory of *déblais* and *remblais*, XI, 417—20.

Maximum Indicator: VI, 83.

Maxwell, J. C.: contour lines, IV, 609; quartic and quintic surfaces, VII, 246, 252; on Cayley, VIII, xx.

Mean Motion (*see* Motion).

Mechanical Construction: curve tracing, VIII, 179—80, XIII, 515—6; of Cartesians, IX, 317, 535—6; of conformable figures, X, 406.

Mechanics: variation of arbitrary constants in, III, 161—5, 200; construction of conformable figures, X, 406; integrator, XI, 52—4; and time, XI, 444; function in, XI, 522—3; curve tracing, XIII, 515—6.

Mehler, F. G.: attractions of polyhedra, IX, 266.

Memoirs: list of, on theoretical dynamics, IV, 584—593; (*see also Ency. Brit. and* the word desired).

Mention, M.: in-and-circumscribed polygon, IV, 294, 303—5.

Mercator's Projection: VIII, 567, IX, 237—40.

Merrifield, C. W.: letter to, VIII, 517—8.

Mersenne, M.: the cycloid, XI, 447.

Metageometry (*see* Hypergeometry, Prepotentials).

Method of Derivations: Arbogast's, II, 257, IV, 265—71, 272—5, 609, XI, 55; binomial theorem and factorials, VIII, 463—73.

Methyl: trees of, IX, 544—5.

Metrical Geometry: II, 592.

Metrical Theory: pure analytical geometry, XI, 556—7; solid geometry, XI, 570.

Meunier, J. B. M. C.: theorem of, III, 38.

Meyer, F.: history of quantics, VIII, xxxi; orthomorphosis, XIII, 175, 187.

Mill, J. S.: logic and mathematics, XI, 432—4.

Miller, W. J. C.: triangles, V, 566; conics, V, 582; negative pedals of ellipsoid and ellipse, X, 576—7; geometrical interpretation, X, 604.

Milner's Lamp: differential equation and construction of, XIII, 3—5.

Minima (*see* Maxima and Minima).

Minimal Surfaces (*see* Surfaces).

Minor: the term, XI, 496.

Minor Function: in Abel's theorem, XII, 111.

Mirrors: systems of rays, XII, 571—5.

Möbius, A. F.: geometry of position, I, 360; reciprocal figures, I, 415; developable from quintic curve, I, 500; tortuous curves, I, 500; circular relation, III, 118—9, IX, 612—7, XI, 188; cubic curves, IV, 120, XI, 479; in-and-circumscribed triangle, IV, 439—41; cubic curves and cones, V, 401, 551; opposite curves, V, 468; equilibrium of four forces, V, 540—1; coordinates of a line, VII, 66, 93; multiple algebra, XII, 472, 473.

Models: Plücker's, of quartic surfaces, VII, 298—302; Wiener's, of cubic surface, VIII, 366—84.

Modular: the term, XIII, 559.

Modular Equations: in elliptic functions, IX, 117—8, 126—37, 169—75, XII, 507—34; errors in Sohnke's paper, IX, 543; for cubic transformation, XII, 46; quintic transformation, XII, 416.

Modular Functions: system of symbols, IV, 484—9; $\chi\omega$, XIII, 338—41.

Modulus: of transformation, the term, IV, 605; table for any prime or composite, VI, 83—6.

Moment: non-Euclidian geometry, XIII, 481—9.

Moment of Inertia: of solid body, IV, 478—80, 559—66.

Monge, G.: transformation of coordinates, IV, 557; theory of *déblais* and *remblais*, XI, 417—20; descriptive geometry, XI, 448—9; reciprocal polars, XI, 465; biographical notice, XI, 586—8; non-Euclidian plane geometry, XII, 221; differential equation of conic, XII, 393.

Monodromic: the term, XII, 432.

Monogenic Function: XI, 80, 537.

Monoid Surfaces: and curves in space, V, 8, 552; and quintic curves in space, V, 24—30, 552, 553, 613.

Monotropic: the term, XII, 432.

Monro, C. J.: flexure of spaces, X, 331—2.

Montucla, J. F.: on Wallis, XI, 642.

Moon: secular acceleration of mean motion, III, 522—61, 568; (*see also* Lunar Theory, Solar Eclipse).

Morley, F.: topology of chessboard, X, 609—10; systems of circles and spheres, XIII, 13.

Motion: of solid body, I, 28—35, 583; secular acceleration of moon's, III, 522—61, 568; Lagrange's equations of, IX, 198—200; on three-bar, IX, 551—80, 585, XIII, 505—16; Sylvester on recent discoveries in mechanical conversion, XIII, 44; (*see also* Dynamics, Elliptic Motion, Kinematics, Lunar Theory).

Motte, A.: problem of tactions, XIII, 151.

Moulton, J. F.: matrices, XI, 256.

Mountains: altitude, and roots of algebraic equation, XIII, 33—7.

Mourey, C. V.: imaginaries, XII, 468.

Mousetrap: the game of, III, 8, X, 256—8.

Moutard, T.: quartic surfaces, VII, 246, VIII, 262.

Muir, T.: history of determinants, I, 581; problem of arrangements, X, 249—51; elimination, XIII, 545—7.

Multiform Series: defined, IV, 456.

Multilinear Operator of MacMahon: XIII, 399.

Multiple Algebra: on, XI, 446, XII, 60—71, 459—89; associative imaginaries, XII, 105—6.

Multiple Integrals: and attractions, I, 5—12, 13—8, 195—203, 204—6, 285—9, 438—44, 581, 586, II, 35—9; demonstration of a theorem of Boole, I, 384—7, 588.

Multiple Sines: X, 1—2.

Multiple Theta-functions (*see* Theta-functions).

Multiplication: of elliptic functions, I, 534—9, 568—76, 589, IX, 138—47, XII, 507; of determinants,

XI, 495; of extraordinaries, XII, 461—2; complex, in elliptic functions, XII, 556—7; (*see also* Transformation).

Multiplier: Jacobi's theory of, I, 276, 279; theory of, in differential equations, X, 102—6; in elliptic integrals, X, 139.

Multiplier Equations: in elliptic functions, IX, 138—47, XII, 507.

Murdoch, P.: *Newtoni Genesis Curvarum per Umbras*, V, 284, 288; curve classification, V, 354; cubic curves and cones, V, 402; the simplex cubical parabola, VI, 101.

Murphy, H.: four points in plane or space, VII, 585.

Murphy, R.: Legendre's coefficients, I, 376.

Music and Mathematics: Sylvester on, XIII, 45.

Napier, J.: logarithms, XI, 447.

Natani, L.: Pfaffian equations, IV, 515.

Nature: notice on Sylvester, XIII, 43—8.

Neg: the abbreviation in groups, XIII, 119.

Negative: the rule of signs, IV, 595—6, XI, 492.

Negative Deficiency: VIII, 397.

Neptune: Newcomb's astronomical work, IX, 180—4.

Neutral: the term, VI, 101.

Newcomb, S.: astronomical work of, IX, 176—84.

Newton, Sir I.: cubic curves, IV, 122; rectilinear motion, IV, 515, 590; central forces, IV, 515, 590; parabolas, V, 284; curve classification, V, 354, 364—6, 396—9; diameter, V, 362; cubic curves, V, 401, 551, XI, 464; conics, V, 562; forms of cubical parabolas, VI, 101; theorem as to roots of equations, X, 5; *Principia*, XI, 447—8; branches of curves, XI, 477; theory of equations, XI, 500, 502; roots of algebraic equation, XIII, 35; Sylvester's work at rule of, XIII, 46; tactions, XIII, 151—69; Newton-circle, the term, XIII, 152.

Newton-Fourier Theorem: imaginary roots, X, 405—6, XI, 143, XIII, 36; extension to complex variables, X, 405—6; theory of equations, XI, 114—21, 122.

Nexal: the term, VIII, 73.

Nil: the term, XII, 66.

Nilfactum: and quantic, VI, 466.

Nilpotent: the term, XII, 61.

Nilvalent: the word, IX, 202.

Nine-point Circle: XIII, 517—9, 520—1, 548—51.

Nitrogen: tree of, IX, 430.

Nivellators: Sylvester's theory of, XIII, 47.

Nodal: the term, VII, 244.

Nodal Anallagmatic: the term, VIII, 67.

Nodal Bicircular Quartic: mechanical description, VII, 182—8.

Nodal Cubic: VI, 171—4, VII, 555.

Nodal Curves: of developable from quartic, V, 135—7; of cubic surface, VI, 450; centro-surface of ellipsoid, VIII, 332—52.

Nodal Director: the term, V, 169—70.

Nodal Generator of Scrolls: V, 169—70, 179—81.

Nodal Isochronic: the term, VII, 473.

Nodal Locus: in singular solutions, VIII, 533.

Nodal Quartic: defined, V, 10; mechanical description of bicircular, VII, 182—8; and hyperelliptic functions, XII, 196—208.

Nodal Quartic Surfaces (*see* Quartic Surfaces).

C. XIV.   15

Nodal Residue of Scrolls: v, 169—70, 181—3, 184, 187.

Nodal Total of Scrolls: v, 169—70, 183—9.

Node-couple: defined, II, 29, IV, 22, XI, 227; curve, and plane, and torse, VI, 355, 582—5; torse, VI, 601.

Node-cusp: v, 265—6, 618.

Node-form: the term, VII, 274.

Nodes: the term, II, 28, IV, 22, 27, 181, v, 295, XI, 468; elimination of, in three bodies, IV, 551, v, 23; number on quartic surface, VII, 133—81; quartic surface with twelve, XIII, 1—2.

Node-tangent: defined, II, 29—32.

Node-triplet: the term, II, 30.

Nodo-focus: of bicircular quartic, VI, 522—3, 523—6; the term, IX, 264.

Non-commutative Algebra (*see* Algebra).

Non-Euclidian Geometry, Memoir on: introduction, XIII, 480—1; geometrical notions, XIII, 481—9; point, line, and plane coordinates, general formulæ, XIII, 489—91; the absolute, XIII, 491—5; distance of a point and line, XIII, 495—7; distance of a plane and line, XIII, 497; theory of two lines, XIII, 497—504.

Non-Euclidian Geometry: VIII, 409—13, XII, 220—38; (*see also* Hypergeometry).

Non-facultative Space: VI, 156.

Non-scrolar Surfaces: quartic and quintic, VII, 245.

Non-unitariants: the term, XIII, 265.

Non-unitary Symmetric Functions: and seminvariants, XII, 239, 275, XIII, 267—98; tables, XII, 273—4.

Norm: and polyzomal curves, VI, 474, 573—5.

Normal Elementary Integral: of differential equation, XII, 396—7, 444.

Normals: in *Ency. Brit.*, XI, 564—5; (*see also* Conics).

Normal Variables: in dynamics, IX, 111.

Notation: algebraic functions, II, 185—8; matrices, II, 185—8; quantics, II, 223; for disturbing function compared, III, 310—8; quantics and abstract geometry, VI, 464—6; differential equations, X, 95—7; for double theta functions, X, 497; theta functions, XI, 47—9, 243—5; umbral, XIII, 301—6.

Nöther, M.: curves in space, v, 613—7; rational transformation, VII, 255; deficiency of surfaces, VIII, 395; sextic curve, IX, 504—7; classification of curves, XI, 451; Abelian function, XII, 149.

Novel-reading: at Cambridge by Cayley, VIII, x—xi, xxiii.

Nullity: Sylvester's theory of, XIII, 47.

Number: time, and space, v, 292, 620, XI, 442—4; theory of equations, XI, 502.

Numbers: a theorem of Lejeune-Dirichlet's, II, 47—8; tables of binary cubic forms, VIII, 51—64; use of Bernoulli's, in analysis, IX, 259—62; arrangements of, X, 570; Sylvester and Hammond on Hamiltonian, XIII, 48; (*see also* Partition of Numbers).

Numbers, Theory of, in *Ency. Brit.*: XI, 592—616; ordinary and complex theories, XI, 592—3; ordinary theory, XI, 594—609, 615—6; theory of forms, XI, 604—9; complex theories, XI, 609—16.

Numbers, Theory of: Pellian equation, IV, 40—2, IX, 477—8, XI, 615, XIII, 430—67; composition of, IV, 70—1, 78—9; specimen table, VI, 83—6; $x^p - 1 = 0$, trisection and quartisection, XI, 84—96; $x^p - 1 = 0$, and quinquisection, XI, 314—6, XII, 72—3; H. J. S. Smith on, XI, 429; imaginaries, XI, 444—5; evolution, XI, 455—6; Wilson's theorem, XII, 45; Sylvester on, XIII, 47; (*see also* Partition of Numbers).

Numerative Geometry: Schubert's, XI, 281—93.

Numerical Equations: X, 3—6.

Numerical Expansions: IV, 470—2.

Numerical Generating Function: X, 339, 408.

Nutation: note on theory of, IX, 194—6.

Obliquity: the term, XIII, 234.

Observations (*see* Orbits, Planet's Orbit, Solar Eclipse).

Octacron: enumeration of polyhedra, V, 38—44.

Octad: the term, I, 586, VII, 133, 152, XII, 590.

Octadic-quartic Surfaces: X, 51.

Octagon: theorem of eight points on a conic, VIII, 92—4.

Octahedral Function: XI, 128—9.

Octahedron: axial system, V, 531—9; automorphic function for, XI, 169, 179—83, 212—6.

Octaves: elliptic functions, I, 127, 586; imaginaries, I, 301.

Octavic Surface: VIII, 401—3.

Octic Function: and Abelian function, XI, 483.

Octics: and twisted cubics, XII, 310.

Octic Surface: on a sibi-reciprocal, X, 252—5; (*see also* Surfaces).

Octo-dianome: the term, VII, 134.

Octo-hexahedron: the term, X, 328.

Odd Branch of Curve: X, 36.

Off-planes: the term, VI, 330, 577, 583—5.

Off-points: the term, VI, 330, 338, 577, 583—5.

Olbers, W.: orbits of asteroids, IX, 177.

Olivier, T.: conics inscribed in quadric surface, I, 557.

Omal: the word, VI, 194, 463, 467—9.

Omega Functions: the term, XI, 453; note on Smith's memoir, XIII, 558—9.

Omphali: the term, VIII, 326.

Operandator: defined, III, 242.

Operations: and substitutions, XIII, 530.

Operators: differential, VII, 8; and seminvariants, XIII, 322—32; MacMahon's multilinear, XIII, 399.

Optics: MacCullagh's theorem in polarized light, IV, 12—20; geometrical construction in, X, 28.

Orbits: Jacobi's canonical formulæ for disturbed motion, III, 76—7; reduction to fixed plane, III, 91—6; variation in plane of planet's, III, 516—8; central forces problem, IV, 516—21; position of, in planetary theory, VII, 541—5; of asteroid, and Newcomb, IX, 176—7; Hamiltonian equations of central, X, 613; Sylvester's work at, XIII, 47; (*see also* Planetary Theory, Planet's Orbit).

Order: of system of equations, I, 457—61, 589; of quantics, defined, II, 221; of curve, II, 569—83, XI, 462; in abstract geometry, defined, VI, 463; of curve and surface, XI, 629.

Ordinary Point for Differential Equations: XII, 394.

Oriani, B.: elliptic motion, III, 474, IV, 528.

Orr, W. McF.: tetrads of circles, XIII, 425.

Orthocentre: the term, XIII, 550.

Orthogonal Surfaces: VIII, 269—91, 292, 569—70; Smith's Prize dissertation, VIII, 558—63.

Orthogonal Surfaces and Curvature, Memoir on: VIII, 292—315; introductory, VIII, 292—3; curvature of surfaces, VIII, 293—300; conormal correspondence of vicinal surfaces, VIII, 301—8; condition that the two surfaces may belong to orthogonal system, VIII, 309—11; family of surfaces, VIII, 312—5.

Orthomorphic Transformation (*see* Orthomorphosis).

Orthomorphosis: of circle into parabola, V, 618, XII, 328—36; of a circle into itself, XIII, 20; general theory, XIII, 170—90; some problems, XIII, 191—205; note on theory, XIII, 418—24; (*see also* Conformal Representation).

Orthotomic: the term, IX, 13.

Orthotomic Circles: and polyzomal curves, VI, 501; and Jacobians, VI, 568.

Orthotomic Curve: of a system of lines in a plane, XIII, 346—7.

Orthotomic Surfaces: in *Ency. Brit.*, XI, 637—8.

Oscnode: defined, II, 28—32.

Oscular: the term, VI, 334, 361, 362.

Ostrogradsky, M. A.: dynamic equations, III, 186, 203; transformation of differential equations, IV, 514; virtual velocities, IX, 207.

Outcrops: the term, VIII, 326, 351.

Oval Chuck for Quartic Curves: VIII, 151—5.

Ovals: of Descartes, I, 479, II, 118, 336, III, 66; and quartic curves, V, 468—70; twice-indented, X, 318; and functions, XI, 540; in *Ency. Brit.*, XI, 549—51; roots of algebraic equations, XIII, 37; Sylvester's work at, XIII, 47; orthomorphosis, XIII, 185—6, 202.

Oxygen: trees of, IX, 427—60.

Π: Wallis's investigation for, XIII, 22—5.

Pagani, G. M.: central forces problem, IV, 520, 590; motion of solid body, IV, 583, 590.

Painvin, L.: last multiplier, IV, 551, 590.

Parabola: inflexions of cubical divergent, V, 284—8, VI, 101—4; classification, V, 356, 395, VI, 101; line and circle, problem, V, 607; polyzomal curves, VI, 542; cubic curves, XI, 478; in *Ency. Brit.*, XI, 548—51, 561—4; orthomorphosis of circle into, XII, 328—36; and epitrochoid, XIII, 86—7; orthomorphosis into circle, XIII, 421—2.

Parabolic Cyclide: IX, 73—8.

Paraboloids: in *Ency. Brit.*, XI, 576—9.

Paradox: the d'Alembert-Carnot geometrical, XII, 305—6.

Paraffins: trees of, IX, 427—60.

Parallel Curves: envelopes and surfaces, IV, 123—33, 152—7, 158—65; and evolutes, VIII, 31—5; theory of, X, 260; the critic in solar eclipses, X, 311—5.

Parallels: and non-Euclidian geometry, XIII, 480—1, 481—9; the terms right and left, XIII, 488, 502.

Parallel Surfaces: of paraboloid, VIII, 7; of ellipsoid, VIII, 9, IX, 591; in *Ency. Brit.*, X, 637—8.

Parametric Class and Order: of systems of cones, V, 552.

Parametric Latitude: VII, 16, IX, 238.

Parametric Relation: VI, 463—4; of triple orthogonal system, VIII, 292—315.

Parazome: the word, VI, 477.

Partial Differential Equations: integral of, III, 166; system of, VIII, 517—8; Jacobi's, in transformation of elliptic functions, XII, 530—3; on a, XIII, 358—61.

Particle: under central forces, X, 575; (*see also* Dynamics).

Partition of Numbers: II, 218, 235—49, V, 48; and quantics, II, 265; supplementary researches, II, 506—12; a problem in, III, 247—9; tactical, V, 294, XI, 443.

Partitions: conjugate, due to Ferrers, II, 419; formulæ in, III, 36—7; problem of double, IV, 166—70; of a close, V, 62—5, 617; problems, VII, 575, X, 611, XI, 61—2; tables, IX, 480—3, XI, 357—64; theorems in trigonometry and, X, 16; in *Ency. Brit.*, XI, 589—91; note on a partition-series, XII, 217—9; non-unitary partition, XII, 273—4; Sylvester's constructive theory of, XIII, 47; of a polygon, XIII, 93—113; and seminvariants, XIII, 269.

Pascal, B.: hexagram of, I, 356; limaçon of, I, 480; some theorems of geometry of position, I, 550—6; lines of, I, 551, 588; curves, XI, 447; inscribed hexagon, XI, 556.

Pascal's Theorem: intersection of curves, I, 25—7; demonstration, I, 43—5; Chasles' form of, I, 45; on, I, 322—8, 414, VI, 129—34, 594; generalized, V, 4; notation of points and lines, VI, 116—22, 594.

Peacock, G.: multiple algebra, XII, 460, 467, 469, 470—1.

Peaucellier, A.: mechanical construction of Cartesian by his cell, IX, 317; cell of, and scalene transformation, IX, 527—34; Sylvester on his discoveries, XIII, 44.

Pedal Curves: Maclaurin on, V, 113—4.

Peirce, B.: orbit of Neptune and Uranus, IX, 180, 182; multiple algebra, XI, 457—8, XII, 60—71; associative algebras, XII, 106, 459, 465; imaginaries, XII, 303.

Pellian Equation: IV, 40—2; tables, IX, 477—80; and theory of numbers, XI, 615; report of committee on, XIII, 430—67.

Pencils: defined, II, 577; homography of, II, 578; intersections of four- and two- lined, V, 484—6; of six lines and cubic curves, VI, 105—15, 593—4.

Pendulum, Spherical: IV, 532—4, 535—7, 541.

Peninvariants: and seminvariants, IV, 241; the term, IV, 606.

Pentagon: a theorem relating to, I, 318—9; Gauss's *Pentagramma Mirificum*, VII, 37—8; Schröter's construction of regular, XII, 47; (*see also* Polygons).

Pentagramma Mirificum: of Gauss, VII, 37—8.

Pentagraph: illustrating a function, XI, 440; curve tracing, XIII, 515—6.

Penultimate Forms: of curves, VIII, 258—61; of surfaces, VIII, 262—3.

Penultimate Quartic Curve: VIII, 526—8.

Penumbral Curve (*see* Solar Eclipse).

Periodic Functions: the term, XI, 529; (*see also* Doubly Periodic Functions).

Periods: of elliptic integrals, IX, 618; of theta functions, X, 467—9.

Permissive Points: in differential equations, XII, 434—41.

Permutants: theory of, II, 16—26, 27; defined, II, 17, IV, 594, 596, 600; Sylvester on, II, 26—7.

Permutations: theory of, I, 423—4; idea of group, II, 124; problem of geometric, V, 493—4; commutants, V, 495—7; colours on faces of polyhedra, V, 539; (*see also* Arrangements, Combinatory Analysis).

Perott, J.: binary quadratic forms, V, 618.

Perpendicular: in non-Euclidian geometry, XIII, 480—1, 481—9.

Perpendicularity: and imaginaries, XII, 466—72.

Perpetuants: and seminvariants, XII, 250—7, XIII, 266; sextic, XII, 257—62; reducible seminvariants, XIII, 308—13; Strohian theory of, XIII, 314—8.

Perspective: of triangles, III, 5; five points in a plane, V, 480—3; theory, XI, 442.

Pfaff, J. F.: problem, IV, 359—63.

Pfaffian Differentials: XIII, 361, 405—14.

Pfaffian Equations: Natani and Clebsch, IV, 515.

Pfaffians: and skew determinants, I, 411, II, 203; the term, I, 589, II, 19, IV, 594, 600; and differential equations, X, 96—7, 106.

Pfaff-invariants: XIII, 405—14.

Physics: relation to mathematics, XI, 444, 449; (*see also* Dynamics, Electricity, Light).

Picard, É.: integral calculus, XII, 408—11.

Pinch-planes: the term, VI, 330, 335, 583—5, X, 53—6.

Pinch-points: the term, VI, 123, 330, 335, 582—5, X, 53—6, XI, 227.

Pineto, S.: tables of logarithms (review), VIII, 95—6.

Pippian: defined, I, 586, II, 381, 397—400, 400—3; and Hessian, II, 383—95; geometrical definition, II, 416.

Pirie, B.: inertia, IV, 564.

Plana, G. A. A.: lunar theory, III, 536, 568, VII, 357—60, 361—6, 367—70, 371—4, 375—6; distribution of electricity, IV, 92, 100—7.

Planar Developables: the term, I, 505.

Planarity: of developables, V, 517.

Plane Curves (*see* Curves, Plane).

Plane-integral: prepotential, IX, 337—43.

Plane of Orbit: variation in, III, 516—8.

Plane Representation of Solids: VII, 26—30.

Planes: diametral, of quadric surface, I, 255—8; point and line defined, II, 561—2; geometry of two dimensions, II, 569—83; MacCullagh's theorem of polar, IV, 12—20; lines and dots of cubic surfaces, VI, 365—6, 373—449; rational transformation, VII, 197—213, 216—9; quadric transformation, VII, 213—6; also lineo-linear, VII, 215—6; determined by point and three lines, VII, 571; fleflec-nodal, of a surface, X, 262—4; kinematics of, XI, 103—10, XIII, 505—16; in *Ency. Brit.*, XI, 571—2; osculating and normal, XI, 579—80; and surface, XI, 629; non-Euclidian geometry, XIII, 481—504; and line distance, XIII, 497.

Planetary Theory: Desboves', III, 185, 203; development of disturbing function, III, 319—43, VII, 511—27; variation in plane of orbit, III, 516—8, VII, 541—5; theorem of Jacobi, III, 519—21; Newcomb's work, IX, 180—4.

Planets: angular distance of two, VII, 377—9.

Planet's Orbit from Three Observations: VII, 384—6, 400—78; introductory, VII, 400—1; the general theory, VII, 401—6; determination of orbit from given trivector, VII, 406—12; time formulæ, Lambert's equation, VII, 412—5; formulæ for transformation between two sets of rectangular axes, VII, 415—7; intersection of orbit plane by single ray, VII, 417—26; trivector and orbit, VII, 426—8; special symmetrical system of three rays, VII, 428—9; Planogram No. 1, meridian 90°—270°, VII, 429, 430—40; No. 2, meridian 0°—180°, VII, 429, 441—51; No. 3, orbit-pole at point *A*, VII, 429, 452—4; No. 4, orbit-pole in ecliptic, VII, 429, 455—9; No. 5, orbit-pole on a separator, VII, 429, 459—67; spherogram and isoparametric lines, VII, 467—8; *e*-spherogram and iseccentric lines, VII, 468—70; time-spherogram and isochronic lines, VII, 470—7.

Planogram: the term, VII, 404; three plates, VII, to face 478; meridian 90°—270°, VII, 429, 430—40; 0°—180°, VII, 429, 441—51; orbit-pole at point *A*, VII, 429, 452—4; in ecliptic, VII, 429, 455—9; on separator, VII, 429, 459—67.

Plates (*see* Diagrams, *also* Tables).

Plato: and geometry, XI, 446.

Playfair, J.: on twelfth axiom of Euclid, XI, 435.

Plerogram: the term, IX, 202.

Plexus: the term, IV, 603, VI, 458; Sylvester's term, XIII, 46.

Plücker, J.: theory of algebraic curves, I, 53, 54; curves and developables, I, 207, 208, 210, 586—7; involution, I, 259, 261; elimination, and theory of curves, I, 344; geometry of position, I, 356, 553—6; geometrical reciprocity, I, 380; reciprocal figures, I, 418; quadric surfaces, I, 421; cubic surfaces, I, 446; transformation of curves, I, 478; singularities of plane curves, I, 586, V, 520—2, 619, XI, 450; cubic curves and cones, IV, 173—8; double tangents, IV, 186; points of six-pointic contact on cubic, IV, 207; cubic curves, IV, 495, 617, V, 402; line geometry, IV, 616—8; hyperboloid coordinates, V, 72; node-cusp, V, 265—6; curve classification, V, 354—400; numbers for singularities of plane curves, V, 424, 476, 517; higher singularities of plane curves, V, 426, 619; pencil intersections, V, 484; numbers of, VI, 68, VIII, 41—5, XI, 469—73; species of cubical parabola, VI, 101; focus, VI, 515, XI, 481; six coordinates of a line, VII, 66; quartic surface models, VII, 298—302; construction of a conic, VII, 592; hypergeometry, VIII, XXXV; theory of curve and torse, VIII, 74, 75—6, 80—1; theory of curves, XI, 467; envelopes, XI, 475—6; note on equations of, XIII, 536.

Pohlke, K.: theorem in axonometry, IX, 508.

Poincaré, H.: lacunary functions, XIII, 415.

Poinsot, L.: polygons and polyhedra, IV, 81—5, 86—7, 609; inertia, IV, 563, 590—1; rotation of solid body, IV, 571—3, 577, 591; kinematics of solid body, IV, 580, 581, 591.

Point: of cubic curve, II, 382; satellite, II, 383; formulæ, II, 405—9; theorems, II, 409—12; plane and line defined, II, 561—2; and ineunt of a curve, II, 574; lattice, III, 40; distances of, from triangle and formulæ, IV, 510—2; tritom, V, 138; the term polar of, V, 570; and abstract geometry, VI, 458; potential of, IX, 278—80; singularities of curves, XI, 468; coordinates of, as functions of parameter, XII, 290—1; and line distance, XIII, 495—7; two-way, XIII, 507; for-forwards, and back-backwards, XIII, 510.

Point-pairs: the term, II, 564—5, VI, 202, 206—7, 208, 210, 269, 594—5; degenerate forms of curves, XI, 218.

Points: distances of, I, 1—4, 581; some theorems in geometry of position, I, 317—28; of inflexion, I, 345—9, 354; of osculation, I, 349—51; harmonic relation of two, II, 96—7; of cesser, defined, IV, 130; critical defined, IV, 130; five in a plane, V, 480—3; correspondence on plane curve of, V, 542—5; and circle, problem, V, 560; correspondence of two on a curve, VI, 9—13, 264—8, VII, 39; notation of, in Pascal's theorem, VI, 116—23; abstract geometry, VI, 463; consecutive, VI, 467—9; system of 16, and polyzomal curves, VI, 501—3, 504—5; problem of random, VII, 585; problem and solution of four in plane or space, VII, 585; four and conic, VII, 587; on particular sextic curve, IX, 504—7; branch- and cross-, X, 317; and lines, problem and solution, X, 570; on a circle, function of, XI, 130; double- and pinch-, XI, 227; Mill on, XI, 432—3; representation on plane, XI, 442; evolution theory of curves, XI, 450—1; at infinity, XI, 464; relation between the distance of five in space, XII, 581—3; analytical formulæ in regard to octad of, XII, 590—3; Sylvester's facultative, XIII, 46; non-existence of a special group, XIII, 212; syzygetic relations, XIII, 224—7; non-Euclidian geometry, XIII, 480—504; coordinates of, and non-Euclidian geometry, XIII, 489—91; (*see also* Orthomorphosis).

Point-systems: and one-dimensional geometry, II, 563—9, 583—86; and two-dimensional geometry, II, 569—83, 586—92.

Poisson, S. D.: attraction of ellipsoids, III, 155; planetary theory, III, 159, 201; variation of arbitrary constants in mechanical problems, III, 163—5, 200, 201, 202; coefficient (a, b) of, III, 163; Hamilton's method of dynamics, III, 173—4, 200; integration of differential equations, III, 180; distribution of electricity, IV, 92—5, 100—7, X, 299, XI, 1; elliptic motion, IV, 522; relative motion, IV, 535, 591; motion of projectile, IV, 541, 591; inertia, IV, 563, 591; rotation of solid body, IV, 566, 569, 573, 591; rotation round fixed point, IV, 582, 591; motion of solid body, IV, 583, 591; attraction of ellipsoidal shell, IX, 302; Jacobi's theorem, X, 108—9, 110—3.

Polar: of point, V, 570, X, 54, XI, 465.

Polar Conjugate: of curve of third class, II, 383.

Polar Reciprocal: I, 230, 378, 416.

Polarization: MacCullagh's theorem, IV, 12—20.

Poles: conjugate, of cubic curve, II, 382—5; two-dimensional geometry, II, 579—83, 586—92; the term, XI, 465.

Pollock, Sir F.: on circumscribed triangle, III, 29—34.

Poloid Curve: IV, 571—2.

Polyacra: triangle-faced, and enumeration of polyhedra, V, 38—44.

Polygons: in-and-circumscribed, II, 87—9, 91—2, 138—44, 145—9, IV, 292—308, 435—41, V, 21—2, VIII, 14—21, 212; partitions of close-, V, 62—5, 617; and triangles, problem, V, 589; potential of, IX, 266—80; automorphic function for, XI, 169, 179—83, 212—6; partitions of, XIII, 93—113.

Polyhedra: Poinsot's four new regular solids, IV, 81—5, 86—7, 609; the problem of, IV, 182—5, 609; autopolar, IV, 185; enumeration of, and triangle-faced polyacra, V, 38—40; partitions of close-, V, 62—5, 617; axial properties, V, 529—39; potential of, IX, 266—80.

Polyhedral Functions (*see* Hypergeometric Series, Schwarzian Derivative).

Polyzomal Curves, Memoir on: VI, 470—576, VII, 115; introductory, VI, 470—2; Part I, polyzomal curves in general, VI, 473—97; definitions and preliminary remarks, VI, 473—4; the branches, VI, 474—6; points common to two branches, VI, 476—8; singularities of a ν zomal, VI, 478—9; zomals with common point or points, VI, 479—81; depression of order of ν zomal curve from ideal factor of branch or branches, VI, 481—5; the trizomal and tetrazomal, VI, 485; intersection of two ν zomals having same zomal curve, VI, 486—7; theorem of decomposition of tetrazomal, VI, 487—9; application to trizomal, VI, 489—94; tetrazomal curve, VI, 494; variable zomal of trizomal curve, resumed, VI, 494—7; Part II, subsidiary investigations, VI, 497—515; preliminary remarks, VI, 497—8; circular points at infinity; rectangular and circular coordinates, VI, 498—9; antipoints; definition and fundamental properties, VI, 499—500; antipoints of circle, VI, 500; antipoints and pair of orthotomic circles, VI, 500; forms of equation of circle, VI, 501; system of 16 points, VI, 501—3; property in regard to four confocal conics, VI, 503—4; system of sixteen points, the axial case, VI, 504—5; involution of four circles, VI, 505—8; locus connected with foregoing, VI, 508—9; formulæ of two sets each of four concyclic points, VI, 509—11; ditto further properties, VI, 512—5; Part III, theory of foci, VI, 515—34; the general theory, VI, 515—7; foci of conics, VI, 517—9; variable zomal applied to conic, VI, 519—21; foci of circular cubic and bicircular quartic, VI, 521—2; centre of circular cubic, and nodo-foci, etc., of bicircular quartic, VI, 522—3; circular cubic and bicircular quartic; symmetrical case, VI, 523; ditto, singular forms, VI, 523—6; analytical theory for circular cubic, VI, 526—8; ditto, for bicircular quartic, VI, 528—30; property that points of contact of tangents from pair of concyclic foci lie in a circle, VI, 530—34; Part IV, trizomal and tetrazomal curves where the zomals are circles, VI, 534—66; the trizomal curve-tangents at I, J, etc., VI, 534—7; foci of conic represented by equation in areal coordinates, VI, 537; theorem of variable zomal, VI, 539—41; relation between conic and circle, VI, 541—2; case of double contact, Casey's equation in problem of tactions, VI, 543; intersections of conic and orthotomic circle on set of four concyclic foci, VI, 543—4; construction of symmetrical curve, VI, 544—6; focal formulæ for general curve, VI, 547; circular cubic, VI, 548—9; focal formulæ for symmetrical curve, VI, 549; symmetrical circular cubic, VI, 549—50; general ditto, VI, 550—3; transformation to new set of concyclic foci, VI, 553; tetrazomal curve, decomposable or indecomposable, VI, 553—4; cases of indecomposable, VI, 554—5; ditto, centres being in line, VI, 555—6; the decomposable curve, VI, 556—7; ditto, centres not in a line, VI, 557—61; ditto, centres in a line, VI, 561—5; ditto, transformation to a different set of concyclic foci, VI, 565—6; theory of Jacobian, VI, 566—8; Casey's theorem for circle touching three given circles, VI, 568—73; a norm when the centres are in line, VI, 573—5; trizomal curves with cusp or two nodes, VI, 575—6.

Poncelet, J. V.: harmonic relations, II, 96; porism of in-and-circumscribed triangle, III, 80—5; rectangular hyperbola, III, 254; in-and-circumscribed polygon, V, 21—2; reciprocal polars, XI, 466.

Pontécoulant, G. de: *Système du Monde*, III, 309—10; *Lunar Theory*, III, 521, VII, 357.

Porism: homographic, defined, III, 74, 84; allographic defined, III, 75, 85; of polygon and correspondence, IX, 94.

Porism of in-and-circumscribed Polygon: II, 87—9, 91—2, 97, 138—44, 145—9, IV, 292—308, VIII, 14—21, 212.

Porism of in-and-circumscribed Triangle: II, 56, 87—90, 91, III, 67—75, 80—5, 229—41, V, 549—50, 579, VIII, 212—57.

Portraits of Cayley: frontispiece to vols. VI, VII, XI.

Pos: the abbreviation in groups, XIII, 119.

Positive: the rule of signs, IV, 595—6, XI, 492.

Postulandum of Curve: the term, I, 583, VII, 140, XII, 501; and capacity, XIII, 115.

Postulation: the term, I, 583, VII, 140, 225, VIII, 394; of curve, XII, 501.

Potential: and attractions, I, 195.

Potentials: of polygons and polyhedra, IX, 266—80; of ellipse and circle, IX, 281—301; Smith's Prize question on, XI, 261—4.

Potential-solid: prepotential, IX, 346—7.

Potential-surface: prepotential, IX, 343—6.

Potenzkreis of Steiner: III, 113.

Power: of a matrix, II, 492—4 ; of homographic function, XI, 253—7.

Power-enders: the term, XIII, 267, 270, 295 ; and reciprocants, XIII, 333.

Powers: successive, of homographic function, X, 305—6, 307—9 ; of roots of algebraical equations, XII, 33—4.

Precession: note on theory of, IX, 194—6.

Prepotentials, Memoir on: IX, 318—423 ; introductory, IX, 318—21 ; prepotential plane, theorem A, IX, 319, 337—43 ; potential surface, theorem C, IX, 320, 343—6 ; potential solid, theorem D, IX, 320, 346—7 ; the prepotential surface integral, IX, 321—30 ; its continuity, IX, 330—4 ; potential solid integral, IX, 334—7 ; examples of foregoing, IX, 347—50 ; surface and volume of sphere, IX, 351—2 ; integral, IX, 352—9 ; prepotentials of uniform spherical shell and solid sphere, IX, 359—79 ; examples, theorem A, IX, 379—93 ; Green's integration of prepotential equation, IX, 393—404 ; examples, theorem C, IX, 404—7 ; examples, theorem D, IX, 407—8 ; prepotentials of homaloids, IX, 408—9 ; Gauss-Jacobi theory of epispheric integrals, IX, 410—7 ; methods of Lejeune-Dirichlet and Boole, IX, 417—23.

Prepotentials: Smith's Prize question, XI, 261.

Presidential Address: to British Association, XI, 429—59.

Prime Numbers: B.A. report on tables of, IX, 462—70.

Prime Roots: tables, IX, 471—7.

Prn: the abbreviation for tortuous curves, XIII, 253.

Principal System of Sextic Curve: VII, 236—8.

Principia: solution of problem, Book I, Sect. V, Lemma XXVII, VII, 30 ; (*see also* Newton).

Principiants: and reciprocants, XIII, 388—98.

Probabilities: questions in theory of, II, 103—4, 594—8, V, 80—5, X, 588, 600—1, 614.

Problems: mechanical, III, 78—9 ; a class of dynamical, continuous impact, IV, 7—11 ; (*see also* Dynamics, Smith's Prize Papers, Three Bodies).

Problems and Solutions from the Educational Times: V, 560—612 ; table of contents, V, 612 ; VII, 546—607 ; table of contents, VII, 607—8 ; X, 566—614 ; table of contents, X, 615—6 ; (*see also* Smith's Prize Papers).

Product: resolvent, IV, 309—13.

Product-theorem: for theta functions, X, 464, 474, 509—46.

Progress of Theoretical Dynamics (*see* Dynamics).

Prohessians: defined, V, 267 ; and developables, V, 513—4.

Prohibitive Points: in differential equations, XII, 434—41.

Projectile: effect of resisting medium, IV, 541.

Projection: stereographic, of spherical conic, V, 106—9 ; of ellipsoid, V, 487—8 ; plane representation of solid figure, VII, 26—30 ; stereographic, VII, 397—9, XI, 187—9, 569 ; blank, VII, 482 ; of surface on plane, VIII, 538—9 ; Mercator's, VIII, 567 ; Mercator's, of skew hyperboloid of revolution, IX, 237—40 ; a problem of, IX, 508—18 ; map, XI, 448.

Prolusions: Sylvester's Astronomical, XIII, 47.

Provectant: defined, II, 514.

Provector: defined, II, 514.

Pseudosphere: the term, XII, 220.

Ptolemy: stereographic projection, XI, 448.

Puiseux, V.: algebraic functions, III, 225 ; spherical pendulum, IV, 533, 591 ; motion of a body, IV, 583, 591.

Pyramid (*see* Polygons, Polyhedra).

Quadrangle: in-and-circumscribed, IV, 307—8 ; differential relation between sides, X, 33—5.

Quadratic Equation: roots, V, 160—1 ; and geometrical interpretation, XI, 258—60.

Quadratic Form: composition, I, 532 ; transformation of, into itself, II, 192—201, 215 ; tables, V, 141—56, IX, 480—3, 486—93.

Quadratic Residues: Eisenstein's geometrical proof, III, 39—43.

Quadratic Transformation of Binary Form: VIII, 398—400; (*see also* Transformation).

Quadratics: resultant of three ternary, and invariant of biternary, IV, 349—58.

Quadric Cones: of six given points, V, 4—6; through given points, X, 575.

Quadric Curves: V, 70—2.

Quadric Equations: transformation of two, I, 428—31, III, 129—31; automorphic linear transformation of, II, 497—505; solution by radicals, X, 9; two related, XI, 37.

Quadric Integral: due to Aronhold, XII, 162—9.

Quadricone: the term, VI, 334, 585, VII, 264.

Quadricovariant: of quantic, II, 520; or Hessian, II, 545; the term, IV, 606.

Quadrics: through nine points, I, 425—7; developable from two, I, 486—95; homographic transformation into itself, II, 105—12, 117, 133—7; theorem on surfaces, III, 115—7; equation of differences for, IV, 242; the term, IV, 604; sections of, V, 133—4; through three lines, VII, 177; in hyperdimensional space, IX, 79—83; covariants of, IX, 537—42; envelope of family of, X, 589; correspondence of confocal Cartesians with right line of a hyperboloid, XII, 587—9; (*see also* Binary Quadrics).

Quadric Seminvariants: generating functions of, XIII, 306.

Quadric Surfaces: *n*-dimensional geometry, I, 62; diametral planes of, I, 255—8; centres of similitude, I, 329—31; note, I, 421—2, 589; abstract of memoir by Hesse, I, 425—7; conics inscribed in a, I, 557—63; envelope of certain, VIII, 48—50; inversion, VIII, 67—71; problem, and hypothetical theorems, VIII, 550; and four-dimensional space, IX, 246—9; a system of, X, 269; Jacobian of, X, 568; in *Ency. Brit.*, XI, 576—9, 632; twisted cubics on, XII, 307—10; focals of, XIII, 51—4; reciprocal lines, XIII, 58—9; (*see also* Geodesic Lines).

Quadric Transformation: between planes, VII, 213—5, 219—21, XII, 100—1; (*see also* Transformation).

Quadri-cubic Curves in Space: V, 16.

Quadricuspidal: the word, VII, 51.

Quadrifactions: the term, IX, 426.

Quadrilateral: and ellipse, V, 604; inscribed in bicircular quartic, X, 231—5; inscribable in circle, X, 578.

Quadrinvariant: of binary quartic, first occurrence, I, 93; of quantic, II, 516; the term, IV, 606; of quadriquadric function, XIII, 67—8.

Quadriquadric: the term and kinds, V, 10, VII, 99.

Quadriquadric Curves: in space, V, 17; on, V, 282; sextic torse for cuspidal edge having, X, 68—72; Abel's theorem, XII, 186—9, 292—8, 321—5; and elliptic functions, XII, 292—8, 321—5.

Quadriquadric Function: two invariants of, XIII, 67—8.

Quadriquadric Transformation: between spaces, VII, 229—30; (*see also* Transformation).

Quadrispinal: the term, VII, 65.

Quantics, Introductory Memoir: II, 221—34, 598—601.

Quantics, Second Memoir: II, 250—75; numerical tables, II, 276—81.

Quantics, Third Memoir: II, 310—35.

Quantics, Fourth Memoir: II, 513—26; definitions, II, 513—5; covariants and invariants of degrees, two, three, four, II, 515—20; calculation of discriminant, II, 520—2; the catalecticant, lambdaic, and canonisant, II, 522—3; bezoutiants, cobezoutiants, II, 524—6.

Quantics, Fifth Memoir: II, 527—57, 604—6; the single quadric, II, 527—9; two or more, and theories of harmonic relation and involution, II, 529—40; cubics, II, 540—5; quartics, II, 545—56.

Quantics, Sixth Memoir: analytical theory of binary and ternary, II, 561—83; general theory of distance, II, 583—92; its style, VIII, xxvii.

Quantics, Seventh Memoir: chiefly ternary cubics, IV, 325—41; tables, IV, 333—41.

Quantics, Eighth Memoir: VI, 147—90; introductory, IV, 147—8; binary quintic, covariants and syzygies of degree 6, VI, 148—53; formulæ for canonical form, VI, 153—4; 18-thic invariant, VI,

154—6; character of equation, auxiliars, facultative and non-facultative space, VI, 156—8; application to quartic equation, VI, 158—61; characters of quintic equation, VI, 161—5; Tschirnhausen's transformation, VI, 165—9; Hermite's application of Tschirnhausen's transformation to quintic, VI, 170; nodal cubic, VI, 171—4; Hermite's criteria, VI, 174—6; his canonical form of quintic, VI, 177—83; imaginary linear transformations, VI, 183—6; application to auxiliars of quintic, VI, 186—7; theorem of binary quantic, VI, 187—90; the binary quintic and sextic, VI, 190.

Quantics, Ninth Memoir: VII, 334—53, IX, 537—42; introductory, VII, 334—5; theory of number of irreducible covariants, VII, 336—7; new formulæ for number of asyzygetic covariants, VII, 337—40; the 23 fundamental covariants, VII, 341—8; tables, VII, 341—6; Gordan's proof for the complete system of 23, and concomitants of quintic, VII, 348—53.

Quantics, Tenth Memoir: X, 339—400; introductory, X, 339—40; numerical and real generating functions, X, 341—8; table 96, X, 349—55; theory of the canonical form, X, 355—62; table 97, X, 362—9; table 98, X, 370—6; derivatives and tables, X, 377—94; numerical generating functions, N.G.F. of a sextic, X, 394—6; table, X, 397—400.

Quantics: defined, II, 221, IV, 594, 604; resultant of, II, 320; discriminants, II, 320; notation of abstract geometry, VI, 464—6; and nilfactum, VI, 466; character of the ten memoirs, VIII, xxx—xxxi; transformable into each other, VIII, 385—7; eliminant of two, XI, 100—2; Sylvester's work in, XIII, 47; syzygetic relations among the powers of linear, XIII, 224—7; and seminvariants, XIII, 363; (*see also* Binary Quantics, Quadratics).

Quarterly Journal of Pure and Applied Mathematics: VIII, xii.

Quartic Curves: transformation, I, 476—80, 589; special family of, I, 496—9; bitangents of, IV, 342—8, VII, 123—4, X, 244, XI, 221—3, 474; cuspidal defined, V, 10; in space, V, 11—5; and ovals, V, 468—70; triangle in-and-circumscribed to a, V, 489—92; with three double points, V, 550, 553; in connexion with cubic and quintic, problem, VI, 580; problem, V, 596; and sextic torse, VII, 99—100; tricuspidal, problem, VII, 589; mechanical description, VIII, 151—5; a penultimate, VIII, 526—8; construction of bicircular, IX, 13—5; and functions of a single parameter, IX, 315—7; with two odd branches, X, 36—7; bicircular, X, 223—42; triple theta functions, X, 446—54; problem and solution, X, 582—6; trinodal, problem, X, 602; singular tangents of, problem, X, 603; degenerate, XI, 220; with cusp at infinity, XI, 408; forms and classification, XI, 480; circular, XI, 481; ground curve in Abel's theorem, XII, 38, 109—216; bitangents of plane-, XII, 74—94; twisted, XII, 428—31; (*see also* Bicircular, Binary, Binodal, *and* Nodal Quartics).

Quartic Developables: and developable surfaces, V, 268—71; reciprocation of, V, 505—10.

Quartic Equations: conditions for systems of equal roots, II, 467—8; evolution, II, 547; Tschirnhausen's transformation, IV, 368—74, V, 449; Sturmian constants, IV, 473—7; nodal curve of developable from, V, 135—7; and quantics, VI, 158—61; solution of $aU + 6\beta H = 0$, VII, 128—9; roots, VII, 551, X, 575; solution by radicals, X, 10.

Quartic Matrix: Hermite's, XII, 367—72.

Quartics: canonical form, II, 548; equation of differences for, IV, 243, 279; the term, IV, 604; roots of, problem, V, 610; conditions for existence of systems of equal roots, VI, 300—12; and three cubics, problem, VII, 546; reality of roots, problem, X, 608.

Quartic Scroll (*see* Scrolls).

Quartic Seminvariants: XII, 20; generating functions, XIII, 306; and perpetuants, XIII, 316.

Quartic Surfaces, First Memoir: VII, 133—81, 609—10; introductory, VII, 133—4; Jacobian surfaces, VII, 134—6; surface by equating to zero a symmetrical determinant, VII, 136—7; surfaces $F(P, Q) = 0$, etc., VII, 138; nodes of quartic surface, VII, 138—40; number of constants contained in a surface, VII, 140—1; general theory of quartic surface with given nodes, VII, 141—4; Jacobian surface of six given points, VII, 144—5; ditto of seven, or an octad of points, VII, 145—8; the dianodal surface, VII, 148—52; octadic surfaces with 9 or 10 nodes, VII, 152—5; dianomes with 9 or 10 nodes, VII, 155; dianodal curve of 8 points, VII, 155—6; ten nodes, VII, 156; dianodal centres of 9 points, VII, 156; result as to dianomes, VII, 156; the symmetroid, (lineo-linear correspondence of quartic surfaces), VII, 157—9; ditto and Jacobian, VII, 160—3; symmetroid with given nodes, VII, 163—6, 259; Jacobian with given lines, VII, 167; correspondence on the Jacobian, VII, 168—70; further investigations as to Jacobian, VII, 171—5; persymmetrical case: Hessian of a cubic, VII, 175; quartics with 11 or more nodes, VII, 176—7; quadric surface through three given lines, VII, 177; condition that five given lines may lie in a cubic surface, VII, 177—8; condition that seven given lines may lie in a quartic, VII, 178; Jacobian of 6 points, VII, 178—9; locus of vertex of quadric cone which touches each of six given lines, VII, 180—1.

Quartic Surfaces, Second Memoir: VII, 256—60, 609—10.

Quartic Surfaces, Third Memoir: VII, 264—97, 609—10; preliminary considerations and classification, VII, 264—7; sextic curves, VII, 267—71; nodal determination, VII, 271—3; quartic surfaces resumed, VII, 273—4; enumeration of the cases, VII, 274—80; notation for cases afterwards considered, VII, 280—1; 16-nodal surface and table, VII, 281—4; 15-nodal surface and table, VII, 285—8; equation of ditto, VII, 288—9; 14-nodal surface and table, VII, 289—92; 13-nodal surface and table, VII, 293—7.

Quartic Surfaces: on, V, 66—9; Steiner, V, 421—3, IX, 1—2; 16-nodal, V, 431—7, VII, 126—7, X, 157—65, 180—3, 604, XII, 95—7; note on, V, 465—7; recent researches, VII, 244—52; Plücker's models, VII, 298—302; some special, VII, 304—13, VIII, 2—11, 25—8; surface and sphere, problem, VII, 589; section of surface, problem, VII, 593; penultimate forms of, VIII, 262—3; symmetrical determinant $= 0$, X, 50—6; 12-nodal, X, 60—2, XIII, 1—2; Hessian of, X, 274—7; tetrahedroid as 16-nodal, X, 437—40; equation of, X, 609; in *Ency. Brit.*, XI, 633—4; (*see also* Cyclide).

Quartic Syzygy: and elliptic integrals, II, 191, IV, 68—9, 609.

Quartic Transformation: of elliptic functions, IX, 103—6.

Quartinvariants: of quantic, II, 516, 520.

Quartisection: theory of numbers, XI, 84—96.

Quasi-inversion: and orthomorphosis, XIII, 192—3.

Quasi-minima: the term, XIII, 42.

Quasi-normal: the term, XIII, 228.

Quaternary: the term, VI, 464.

Quaternary Function: Hessian of, IX, 90—3.

Quaternions: certain results, I, 123—6, 127; algebraic couples, I, 128—31; rotation, I, 405—9, 589, V, 537; formulæ of, II, 107; transformation of quadrics, II, 135; skew determinants, II, 214; transformation of coordinates, IV, 559; the equation $qQ - Qq' = 0$, XII, 300—4, 311—3; matrices, XII, 303; multiple algebra, XII, 474; hydrodynamical equations, XIII, 8; versus coordinates, XIII, 541—4.

Quet, J. A.: relative motion, IV, 536, 592.

Quetelet, M. A.: theory of Gergonne, and on caustics, II, 339; wave surface, IV, 433—4.

Quinquisection: theory of numbers, XI, 314—6, XII, 72—3.

Quintic Curves: and developables, I, 500—6; in space, V, 15—6, 20, 24—30, 552, 553, 613; in connexion with cubic and quartic, V, 580.

Quintic Developables: and surfaces, V, 272—8, 518.

Quintic Equations: conditions for systems of equal roots, II, 468—70; equation of differences, IV, 150—1, 246—61, 276—91; Tschirnhausen's transformation, IV, 375—94; tables, IV, 379—80, 387—90; Jerrard's researches, V, 50—4, 77, 89; character of, VI, 161—5; solvability by radicals, VII, 13—4, X, 11; theorem of Abel, XI, 132—5; solvable case of, XI, 402—4; and elliptic functions, XIII, 473; their sextic resolvents, XIII, 473—9.

Quintic Matrix: XII, 376—80.

Quintics: auxiliary equation for, IV, 309—24; Jerrard's form, IV, 392; soluble by elliptic functions,

IV, 484; the term, IV, 604; MacClintock on, IV, 609—16; theorem of Abel as to soluble, V, 55—61, XI, 402—4; discriminant of, problem, V, 592; conditions for existence of systems of equal roots, VI, 300—12; concomitant system of, X, 342; syzygies among covariants of, X, 346—55; canonical form, X, 355—62; resolvent sextic of, XI, 396; bitangents of the, XIII, 21; and seminvariants, XIII, 363—5; (*see also* Binary Quintics).

Quintic Seminvariants: and perpetuants, XIII, 309.

Quintic Surfaces: and developables, V, 272—8, 518; recent researches, VII, 244—52.

Quintic Transformation: of elliptic functions, IX, 122, 148, XII, 522—5.

Quippian: the word, II, 381, 396—7.

Quotient: G/H in theory of groups, XIII, 336—7.

Raabe, J. L.: summation of series, II, 15.

Radials: the term, XI, 259, XIII, 179.

Radicals: and solvability of equations, VII, 13—4, X, 8—11; theory of equations, XI, 511—20, 521; Galois and theory of, XI, 543; Weierstrassian cubic transformation, XIII, 31.

Radicals (Chemical): number of univalent, IX, 544—5.

Radii: the term link- and bar-, VII, 183; curvature of wave-surface, XIII, 248.

Range: defined, II, 577; homography, II, 578.

Rank of Seminvariants: XII, 22.

Ratio: and abstract geometry, VI, 457—62.

Rational Functions: the term, IV, 603—4.

Rationalisation: of algebraic equations, II, 40—4.

Rational Transformation (*see* Transformation, Rational).

Ray, M. N.: solution of equations, X, 610—11.

Ray Planes: and biaxal crystals, IX, 107—9.

Rayleigh, Lord: fluctuating functions, IX, 19—20.

Rays: special symmetrical system of three, VII, 428—9; the term, X, 55; systems of, XII, 571—5.

Real Generating Function: X, 339.

Real Intersections of Curves: IX, 21.

Reciprocal Figures: I, 415—20.

Reciprocal Matrix: II, 481.

Reciprocal Polars: I, 416, 421; Monge, XI, 465; Poncelet, XI, 466.

Reciprocals: of quartic scrolls, VI, 317—27; of cubic surfaces, VI, 368—455; of quartic surfaces, VII, 305; of centro-surface of ellipsoids, VIII, 363; equation of conic, VIII, 522—3; reciprocal lines, XIII, 58—9.

Reciprocal Surfaces, Theory of: VI, 329—58, 577—81, 582—91, 596—601; extension of Salmon's fundamental equations, VI, 329—31; developments, VI, 331—4; new singularities, VI, 334—41; application to a class of surfaces, VI, 341—2; flecnodal curve, VI, 342; surfaces of revolution, in connexion with spinodal and flecnodal curve, VI, 342—4; flecnodal torse, VI, 345; general surface of order n without singularities, VI, 345—6; formula for β', VI, 347—53; recapitulation, VI, 353—5; addition, VI, 355—8; Zeuthen, VI, 596—601; theory, XI, 225—34.

Reciprocants: of quantic, II, 320; the term, IV, 607, XIII, 366; and sextactic points, V, 618; of cubic, VI, 73; and invariants, XII, 393; Sylvester on, XIII, 47—8; tables of pure, to weight eight, XIII, 333—5; and differential invariants, XIII, 366—404; Halphen on, XIII, 366, 368—81, 381—98; Cockle, XIII, 366, 367—8; Ampère and Lagrange, XIII, 366; Sylvester, XIII, 366, 379—81, 381—98; MacMahon, XIII, 399—404.

Reciprocation of Quartic Developable: V, 505—10.

Reciprocity: geometrical, I, 377—82; and quantics, II, 232, 234; law of, for invariants, II, 516; and homography, II, 578.

Rectangle : potential of, IX, 278—80.

Rectilinear Motion : IV, 515—6.

Reduced Latitude : VII, 16, IX, 238.

Reducible Seminvariants : and perpetuants, XIII, 308—13.

Reducible Syzygies (*see* Syzygies).

Reduction : of transcendental integrals, X, 214—22.

Reech, F. : contour lines, IV, 609.

Reflection : caustics by, I, 273—5, II, 118—22, **129.**

Region : the term, IX, 331.

Regular : the term, VI, 457, 459.

Regulator : the term, VII, 402.

Regulus : the term, XI, 573, 632.

Rehorovsky, W. : symmetric functions, II, 602.

Relation : and abstract geometry, VI, 457—62; omal, VI, 463; parametric, VI, 463—4; a discriminant, VI, 467; Jacobian, VI, 467.

Relink : the term, V, 521.

Remblais : theory of, XI, 417—20, 449, 587.

Reports : on progress of theoretical dynamics, III, 156—204; on progress in solution of certain problems in dynamics, IV, 513—93; on Pellian equation, XIII, 430—67.

Representation : analytical, of curves in space, IV, 446—55, 490—5, XI, 83; of solid figure in plane, VII, 26—30; of surfaces on a plane, VIII, 538; of variables by correspondence of planes, X, 316—23; conformal, XI, 78—81; graphical, of binodal quartic and the elliptic functions, XIII, 9—19; Sylvester on graphical, XIII, 47; (*see also* Orthomorphosis, Transformation).

Réseau : the term, VII, 253.

Residuation : of cubic curve, IX, 211—4, XII, 115—6; of curves, XII, 502; Sylvester's theory of, XIII, 47.

Residues : Cauchy's theorem on, I, 148, 174; Eisenstein's geometrical proof of quadratic, III, 39—43; nodal, of scrolls, V, 169—70, 181—3, 184, 187.

Resisting Medium : motion in, IV, 541.

Resolvent Equations : sextic, of Jacobi and Kronecker, XIII, 473—9.

Resolvents : after Lagrange, IV, 309; of quintics, XI, 396.

Resultant : the term, I, 63, 337, IV, 597, 602—3, VI, 466—7; of quantics, II, 320; of two equations, II, 440—53, VI, 292—9; of two binary quantics, IV, 1—4, IX, 16—7; of three ternary quadratics, IV, 349—58; of two binary cubics, V, 289; of forces, X, 589.

Resultor : defined, II, 59.

Reversion : of series, IV, 30—7, 54—9.

Reuschle, K. G. : mathematical tables, IX, 468—9, 473, 485, 494—9, XI, 95—6; theory of numbers, XI, 85—6, 315, 612.

Rhamphoid Cusp : V, 265—6, 618.

Rhizic Theory : root-limitation, IX, 34—8.

Ribaucour, C. R. : orthogonal surfaces, VIII, 569—70.

Riccati, J. F. : solution of equation, VII, 9—12.

Richelot, F. J. : Abelian integrals, I, 366, 367; solution of equation $x^{257} - 1 = 0$, I, 564; porism formula, II, 90; in-and-circumscribed triangle, III, 237—41; spherical pendulum, IV, 534, 592; rotation of solid body, IV, 577—8, 592; rotation round fixed point, IV, 583, 592; two quartic curves, X, 584; integral of Euler's differential equation, XIII, 525—9.

Richmond, H. W. : Pascal's theorem, VI, 594.

Riemann, G. F. B. : doubly infinite series, II, 593; genus of curve, V, 476—7, 517; Abelian integrals, V, 521, XI, 30; Abelian functions, VI, 2, 264, 593; elliptic geometry, VIII, xxxvii; transformation and theory of invariants, VIII, 387; surface of, and correspondence, X, 317, 323; bitangents of quartic, XI, 221—3; fractional differentiation, XI, 235—6; notion of space, XI, 435—7; correspondence of points, XI, 440; deficiency of curves, XI, 450; hyperelliptic functions, XI, 454—5; transformation, XI, 482; elliptic functions, XI, 534, 537, 539, 546; theory of numbers, XI, 616; series, XI, 627; minimal surface, XI, 639; Abelian functions and plane quartics, XII, 74, 87, 95; linear differential equations, XII, 396; orthomorphosis, XIII, 180, 189, 204.

Right-handed Root-limitation: IX, 22—3.

Roberts, C. A.: Pellian equation, XIII, 467.

Roberts, M.: geodesic lines on ellipsoid, VII, 34.

Roberts, S.: description of nodal bicircular quartic, VII, 182—8; points on cubic curve, VII, 549; three-bar curve, IX, 551; symmetrical determinant, X, 579—80; theorems of squares, XI, 294; curves, XI, 481; kinematics of a plane, XIII, 505.

Roberts, W.: transformations of curves, I, 471—5, 478; surface parallel to ellipsoid, IV, 158—65.

Roberts, W. R. W.: cyclide, IX, 75.

Rodrigues, O.: motion of solid body, I, 28—35, 124, 405, 583; quaternions, I, 124, 586; rotation of solid body, I, 237; skew determinants, I, 335; expansions in multiple sines, I, 583; attraction of ellipsoids, III, 149—53; transformation of coordinates, IV, 558, 559, 592, XI, 575; kinematics of solid body, IV, 581, 592; on rotation formulæ, V, 537; correspondence of homographies and rotations, X, 153.

Rohn, K.: quartic surfaces, VII, 609—10; quartic surface with twelve nodes, XIII, 2.

Roof: the term in non-Euclidian geometry, XIII, 484.

Root-limitation: geometrical representation, IX, 21—39; general theory, IX, 22—7; intercalation theory for right line, IX, 28—33; rhizic theory, IX, 34—8.

Roots: in forms called trees, III, 243, XI, 365—7; of algebraic equations, IV, 116—9, XI, 506—21, XII, 33—4, XIII, 33—7; quadratic equations, V, 160—1; of equations, Cauchy's theorem, IX, 21—39; of unity, IX, 263; of quartic, reality of, X, 608; imaginary, of equations, XI, 114—21, 502—6; equal, of equations, XI, 405—7; theory of real equations, XI, 497—502; Sylvester's work at, XIII, 46; ninth, of unity, XIII, 66; Waring's formula for sum of mth powers of, of an equation, XIII, 213—6; of a quantic, symmetric functions of, XIII, 271—85.

Roots of Unity: prime, XI, 56—60.

Rosenhain, J. G.: theta functions, VIII, xlii, X, 464, 499; double theta functions and 16-nodal quartic surface, X, 158, 162; theory of numbers, XI, 60; double theta functions, XI, 454.

Rotation: of solid body, I, 28—35, 237, 583, II, 107, III, 475—504, IV, 566—80, 592; quaternions and theory of, I, 405—9, 589; formula of, I, 586; and elliptic motion, III, 475; of earth, III, 485; infinitesimal, V, 498—9, VI, 24—6; of group of polyhedra, V, 529, 559; Euler's memoir of 1758, VI, 135—46; equilibrium of, VII, 91—5; and homography, X, 153—4; in conformal representation, XI, 78.

Roulette: the term, XI, 447.

Route: the term, XII, 640.

Rowe, R. C.: memoir on Abel's theorem, XI, 29—36; partitions of a polygon, XIII, 93, 112.

Royal Society: Cayley elected a fellow in 1852, VIII, xiii; Croonian lecture founded, VIII, xv; medals bestowed on Cayley, VIII, xxi.

Rudio, F.: inverse centro-surfaces, XII, 457—8.

Rueb, A. S.: motion of solid body, I, 464; spherical pendulum, IV, 534; rotation of solid body, IV, 573—4, 592.

Rule of Signs: and determinants, XI, 492.

Sabine, Sir E.: death of, XI, 429—30.

Sadleir, Lady Mary: endowments by, VIII, xv.

Sadlerian Professorship: Cayley appointed to, VIII, xvi.

Safford, T. H. : orbits of Neptune and Uranus, ix, 183.

St Laurent, M. : on caustics, ii, 118, 121, 122, 347, 355, 368.

Salmon, G. : cubic surfaces and triple tangents, i, 446, 456, 589 ; linear transformations and elimination, i, 457—61 ; singular contact, i, 486 ; curves and developables, i, 492, 587 ; developable
from quintic curve, i, 500—1, 505 ; systems of equations, i, 533 ; geometry of position, i, 555 ;
hyperdeterminants, i, 579, ii, 598—601 ; on a plane touching a surface, ii, 29 ; triple tangent
planes of third order, ii, 29 ; invariant of ternary cubic, ii, 325 ; quippian, ii, 381 ; tables
of covariants, ii, 536—7 ; binary quartics, ii, 549 ; tangential of cubic, ii, 558 ; equation of orthotomic
circle, iii, 48—50 ; reciprocal surfaces, iv, 21—7, vi, 329—58, 359, 582—91 ; surface parallel to
ellipsoid, iv, 158—65 ; double tangents, iv, 187—206, 343, xi, 473—4 ; cubic curves, iv, 188 ;
conics and five-pointic contact, iv, 207—39 ; higher algebra, iv, 608 ; curves in space, v, 9—20,
614 ; quartic surfaces, v, 66, vii, 136 ; cubic surfaces, v, 140, vi, 359 ; scrolls, v, 168—9, 193,
200 ; prohessian, v, 267 ; involution, v, 301 ; higher singularities of plane curves, v, 620 ; plane
curves, vi, 2 ; invariants, vi, 108 ; quintics, vi, 154 ; hyperspace, vi, 191 ; elimination, and curves
which satisfy given conditions, vi, 192 ; extension of his fundamental equations, vi, 329—31 ;
polyzomal curves, vi, 472, 531, 560 ; tetrahedral scrolls, vii, 52, 65 ; sextic torse, vii, 113, 114 ;
centro-surface of ellipsoid, vii, 130, viii, 316, 320, 323 ; rational transformation between two spaces,
vii, 226, 237 ; bicircular quartic, vii, 575 ; locus *in plano*, vii, 606 ; correspondence with Cayley,
viii, xv ; on Cayley, viii, xxv ; evolutes and parallel curves, viii, 33 ; theory of curve and torse,
viii, 72, 76—9, 87—91 ; theory of invariants, viii, 386 ; transformation of unicursal surfaces, viii,
390, 391 ; residuation, ix, 211 ; triple theta functions, x, 444 ; tortuous curves, xi, 9 ; higher
plane curves, xi, 217 ; Gaussian theory of surfaces, xi, 332 ; concomitants of ternary cubic, xi,
342 ; tables for binary sextic, xi, 377 ; Jacobian sextic equation, xi, 390, 400 ; equal roots of
equations, xi, 407 ; works on geometry, xi, 546 ; minimal surfaces, xi, 639 ; bitangents of quintics,
xiii, 21 ; wave surfaces, xiii, 252.

Satellite Line : ii, 383, v, 359.

Scalars and Quaternions : xiii, 541.

Scalene Transformation of Plane Curve : ix, 527—34.

Schellbach, C. H. : solution of Malfatti's problem, iii, 44—7.

Schläfli, L. : discriminants, i, 584 ; elimination, ii, 181—4, 404 ; symmetric functions, ii, 454 ; hyperdeterminants, ii, 598—601 ; resultants, iv, 2—4 ; numerical expansions, iv, 471 ; cubic surfaces, vi,
359, 361, 362, 372, vii, 250 ; quartic surfaces, vii, 308 ; modular equation for cubic transformation,
xiii, 64—5.

Schlömilch, O. : attractions, i, 288 ; a definite integral, iv, 29.

Schoolgirl Problem : i, 483, 589, v, 95—7.

Schottky, F. : theta functions, xi, 242—9.

Schröter, H. : Steiner's quartic surface, v, 423 ; construction of regular pentagon, xii, 47.

Schubert, H. : elliptic motion, iii, 473, 474, iv, 523 ; *abzählende Geometrie*, xi, 281—93, 459.

Schwarz, H. A. : inverse elliptic functions, i, 586 ; developable surfaces, v, 517—9 ; deficiency, vi, 2 ;
scrolls, vi, 312 ; quintic scrolls, vii, 250, 252 ; projections, ix, 508 ; surface of minimum area, x,
63 ; hypergeometric series, xi, 125 ; orthomorphosis, xii, 328, xiii, 188, 191, 192, 193, 202 ;
Kummer's differential equation, xiii, 69.

Schwarzian Derivative and Polyhedral Functions, Memoir : xi, 148—216 ; introductory, xi, 148—51 ;
Part I, xi, 151—79 ; the derivative, xi, 151—3 ; quadric function of three or more inverts, xi, 153—6 ;
functions P, Q, R, xi, 156—7 ; table ditto, xi, 158—9 ; differential equations involving (x, z) and (s, x),
xi, 160—9 ; Schwarzian theory, xi, 169—76 ; connexion with differential equation for hypergeometric
series, xi, 176—9 ; Part II, the polyhedral functions, xi, 179—216 ; origin and properties, xi,
179—83 ; covariantive formulæ, xi, 184—5 ; the forms of $f\,5$ and $h\,5$, xi, 185—6 ; stereographic
projection, xi, 187—9 ; groups of homographic transformations, xi, 189—90, 196—208 ; the regular

polyhedra, XI, 190—5; system of fifteen circles, XI, 208—12; regular polyhedra as solid figures, XI, 212—6.

Schwarzian Derivative, and Polyhedral Functions: XI, 149, 173, 177; and Kummer's differential equation, XIII, 69; and reciprocants, XIII, 366.

Schwarzian Function: Sylvester on the, XIII, 381.

Scientific Memoirs Catalogue: report on, V, 546—8, 620.

Scrolar: the term, VI, 334.

Scrolls, First Memoir: V, 168—200.

Scrolls, Second Memoir: V, 201—20, VI, 360; degeneracy, V, 201—3; with two directrix lines, V, 203—5; twofold directrix line, V, 205—7; scroll equations, V, 207—10; cubic scrolls, V, 210—3; quartic scrolls, V, 214—9; general theory, V, 219—20.

Scrolls, Third Memoir: VI, 312—28; quartic scrolls, VI, 312—4, 314—7, 328; reciprocals, VI, 317—25, 325—7.

Scrolls: cubic, V, 90—4, 110—2; quartic, VI, 451; tetrahedral, VII, 48—53; on certain, VII, 54—65; recent researches, VII, 250—1; and octic surfaces, X, 79—92; flexure and equilibrium, XI, 317—22; the term, XI, 573; applicable upon a given skew surface, XIII, 231—7.

Secondary Caustics (*see* Caustics).

Secular Acceleration of Moon's Mean Motion: III, 522—561.

Seeber, L.: mathematical tables, IX, 491—2.

Segar, H. W.: development of $(1+n^2x)^{m/n}$, XIII, 354—6.

Segner, J. A. v.: inertia, IV, 561—2, 592.

Segregates: the term, X, 339, 345; table of, X, 349—55.

Semi-cubical Parabola: and Abel's theorem, XII, 180—6.

Seminvariants, memoir: XII, 239—62; introductory, XII, 239—40; multiplication of two symmetric functions, XII, 240—8; capitation and decapitation, XII, 248—50; perpetuants, etc., XII, 250—7; sextic perpetuants and sextic syzygies, syzygants, XII, 257—62.

Seminvariants: the term, IV, 241, 606; theory of, XII, 22—9, 344—57, XIII, 362—5; tables, XII, 275—89, XIII, 217—23; theorem relating to, XII, 326—7; (*see also* Covariants, Invariants).

Seminvariants and Symmetric Functions, Memoir: XIII, 265—332; introductory, XIII, 265—6; the coefficients $(a, b, c, ...)$, XIII, 266—71; symmetric functions of the roots, XIII, 271—85; MacMahon form of equation, XIII, 286—8; the *I-and-F* problem, and solution by square diagrams, XIII, 288—98; MacMahon linkage, XIII, 298—301; umbral notation, Stroh's theory, XIII, 301—6; symmetric functions of a given degree: generating functions, XIII, 306—8; reducible seminvariants—perpetuants, XIII, 308—13; Strohian theory resumed: application to perpetuants, XIII, 314—8; investigation of the values of the foregoing functions, XIII, $\Pi_{10}(x+y)$ $\Pi_{15}(x+y)$ and $\Pi_{10}(x+y+z)$, XIII, 318—21; the operators, XIII, 322—32.

Senate House Problems: IX, 246—9, XI, 265—7, XIII, 538—40; (*see also* Smith's Prize Papers).

Separations of Partitions: II, 603; (*see also* Partitions).

Separator: the term, VII, 402.

Septic, Binary: generating functions of, X, 408—21.

Septic Transformation in Elliptic Functions: IX, 148—52, X, 333—8.

Series: a double infinite, II, 8—10, 593; sums of, III, 124—8, XIII, 49—50; reversion, IV, 30—7; formula for reversion, IV, 54—9; in integration of differential equations, VIII, 458—62; hypergeometric, XI, 17—25, 125—7; in *Ency. Brit.,* XI, 617—27 (finite, XI, 617—20; infinite, XI, 620—7).

Serpoloid Curve: IV, 571—2.

Serret, J. A.: attractions, I, 288; integral calculus, I, 313; elliptic functions, III, 3; integration of differential equations, III, 185—6, 203; problem of two centres, IV, 528, 531, 592; theory of substitutions, VII, 47; curves of curvature, VII, 331, XII, 601—38; orthogonal surfaces, VIII, 279, 292; indefinite integration, IX, 500—3; transformation of cubic function, XI, 411; minimal surfaces, XII, 595.

Serret, P.: syzygetic relations, XIII, 224.

Servois, J. F.: multiple algebra, XII, 468.

Sets: homology of, III, 35.

Sextactic Points: and plane curves, V, 545, VI, 217; and reciprocants, V, 618.

Sextactic Reciprocant: XIII, 387.

Sextic: binary, and quintics, VI, 190; unicursal, VI, 248; and cubic curves in pencil of six lines, VI, 593—4; the anharmonic-ratio, VII, 314—5; bicursal, IX, 551, 581—6; tricircular, IX, 562—70; numerical generating function, X, 394—6.

Sextic Cone: circumscribed to quartic surface, VII, 265; and nodes, table, VII, 291.

Sextic Curves: foci of conics, VII, 1—4; rational transformation, VII, 236—8; and nodes, VII, 256—7; and quartic surfaces, VII, 267—71; mechanical description, VIII, 138—44; with five double points, IX, 504—7.

Sextic Developable: V, 279—83, 511—9, VI, 87—100.

Sextic Function: and Abelian functions, XI, 483.

Sextic Resolvent: Jacobian, IV, 310, XI, 389—401, XII, 493—9.

Sextic Resolvent Equations: of Jacobi and Kronecker, XIII, 473—9.

Sextic Seminvariants: and perpetuants, XIII, 317.

Sextic, Spherical: and oval, V, 469.

Sextic Syzygies: XII, 257—62, 273.

Sextic Torse (*see* Torse, on a certain Sextic, *also* Torses).

Shanks, W.: log 2, XI, 70.

Sharp: the term, VIII, 406—8, XIII, 265, 291, 304—6, 362.

Sharp-cone: the term, VIII, 102.

Sheets: roots in algebraic equations, IV, 116—9; cubic curves, IV, 120—2.

Shell: formulæ for potential of, IX, 266—7; attraction of ellipsoidal, on exterior point, IX, 302—11.

Sibi-reciprocal Surfaces: VI, 21, X, 252—5.

Siebeck, F. H.: binodal quartic and graphical representation of the elliptic functions, XIII, 10.

Signs: rule of, IV, 595—6, XI, 492.

Sign Symbols: theorems, VIII, 535—7.

Simple Cone: defined, V, 402, 404, 551.

Simple Groups: XIII, 533.

Simultaneous Equations: Jacobi's theorem in, XII, 39.

Simultaneous Roots of Two Equations: Jacobi's theorem in, XII, 123—5.

Sines, Multiple: X, 1—2.

Single Theta Functions: memoir, X, 473, 476—97; and double theta functions, X, 155—6, 186—9; and elliptic functions, XI, 250—1; linear transformation, XII, 50—5.

Singular Curve: on surface, VII, 244.

Singularities: of curves and developables, I, 208; of tortuous curves, I, 500; of surfaces, II, 28—32, IV, 21—7, VI, 123—8, 334—41, 354, 582—5, VIII, 394—8, XI, 225—6, 630—1; of plane curves, V, 424—6, 476—7, 520—8, 619, VI, 3, XI, 30—6; of curves and torses, V, 516; compound, V, 525; of curves in space, V, 613; of cubic surfaces, VI, 363; reciprocal surfaces, VI, 596—601; of curves, XI, 486—70.

Singular Point: for differential equations, XII, 395; integrals in domain of, XII, 395—402.

Singular Solutions of Differential Equations: IV, 426—7; of first order, VIII, 529—34 X, 19—24.

Six Coordinates of a Line (*see* Coordinates, Six of a Line).

Six Lines: Sylvester's involution of, VII, 66.

Six-pointic Contact: on cubic, IV, 207.

Sixteen-nodal Quartic Surfaces: I, 587, V, 431—7, VI, 126—7, 281—4, X, 157—65, 180—3, 437—40, 464, 548—51, 604, XII, 95—7.

Skew: the word, I, 332.

Skew Antipoint: the term, IX, 65—6.

Skew Convertible Matrices: II, 489.

Skew Covariants: II, 233.

Skew Cubics: demonstration of Chasles' theorem, I, 212.

Skew Curvature: the term, I, 234.

Skew Determinants: the term, and some properties of, I, 332—6, 410; on, I, 410—3, 589; researches, II, 202—15, IV, 72—3; and transformation, II, 497; a theorem, IV, 72—3.

Skew Hyperboloids: geodesic lines, VIII, 174—8, 188—99; of revolution, projection of, IX, 237—40.

Skew Matrix: IV, 602.

Skew Polars: I, 378.

Skew Reciprocals: the term, I, 415.

Skew Surfaces (*see* Scrolls).

Slope Lines: IV, 108—11, 609.

Smith, H. J. S.: higher singularities of plane curves, V, 619; transformation of elliptic functions, IX, 174—5; report on mathematical tables, IX, 461—99; death of, XI, 429; theory of numbers, XI, 612; theta and omega functions, XII, 50; theta functions, XII, 337; transformation, XIII, 38; predecessor of Sylvester in Oxford chair, XIII, 44; on a memoir by, XIII, 558—9.

Smith, P.: on Lagrange's solution of caustic, II, 353.

Smith's Prize Papers: VIII, 414—35, 436—8, 439—57, 474—90, 496—516, 538—9, 551—5, 558—63; infinitesimal rotation, VI, 24—6; general equation for virtual velocities, IX, 205—8; solutions and remarks, IX, 218—36; Bernoulli's numbers in analysis, IX, 259—62; problems and solutions 1877, X, 39—46; question on theory of equations, XI, 115; on potentials, XI, 261—4.

Sohnke, L. A.: motion in resisting medium, IV, 541; transformation of elliptic functions, IX, 113, 114; ditto table, IX, 128—35; modular equations, IX, 543.

Solar Eclipse, Graphical Construction: VII, 390—1, 479—92; geometrical theory, VII, 392—6, X, 310—5; general explanation, VII, 479—82; modification for single blank projection, VII, 482—4; construction of relative orbits, VII, 484—7; geometrical theory of projection of penumbral curve, VII, 488—9; details and application to eclipse 21—22/12/70, VII, 489—92.

Solid Body: motion of, I, 28—35, 583; geometrical representation of motion, I, 234—6; rotation round fixed point, I, 237—52, 336; rotation of, I, 462—4, III, 475—504, IV, 577, 592; four forces acting on, IX, 201.

Solid Integral Prepotential: IX, 334—7.

Solid of Revolution: attraction of, I, 508.

Solids: Poinsot's four new regular, IV, 81—5, 86—7; plane representation of, VII, 26—30.

Soluble Quintics: IV, 484, V, 55—61, XI, 402—4.

Solutions (*see* Problems, Problems and Solutions, Smith's Prize Papers).

Somoff, J.: rotation of solid body, IV, 577, 592.

Space: of any number of dimensions, and quantics, II, 222; facultative and non-facultative, VI, 156; dimensions and abstract geometry, VI, 456—7; rational transformation, VII, 189—240; multiple, VIII, xxxiii—v; points and lines, correspondence in, VIII, 566; flexure, X, 331—2; the term in five-dimensional geometry, IX, 79; theories of, XI, 434—7; curves in, XI, 489; elliptic, and non-Euclidian geometry, XIII, 481; (*see also* Hyperspace).

Special Conditions for Curves: VI, 193.

Species: of Quartic Scrolls, V, 201, VI, 328; twenty-three, of cubic surfaces, VI, 359—60.

Specific: the term, XIII, 290.

Sphere: powers of, I, 581; and polyhedron, V, 531; problem and solution, VII, 563; prepotentials, IX, 351—2, 359—79; in *Ency. Brit.*, XI, 571—2.

Spherical Conics: theorem, IV, 428; stereographic projection, V, 106—9; (*see also* Polyzomal Curves).

Spherical Curves of Curvature: surfaces with, XII, 601—38.

Spherical Pendulum: IV, 532—4, 535—7, 541.

Spherical Triangle: theorem of, IV, 80; nine-point circle of, XIII, 548—51.

Spherogram: the term, VII, 404; and isoperimetric lines, VII, 467—8; e- and iseccentric lines, VII, 468—70; time- and isochronic lines, VII, 470—7; two plates, VII, to face 478.

Spheroidal Trigonometry: IX, 197.

Spheroid, Oblate: geodesic lines on, VII, 15—25.

Spheroquartic: the term, VIII, 262.

Spinodal Curve: VI, 342—4, 583—5.

Spinode: synonymous with cusp, II, 28, IV, 22, 27; the term, V, 521; plane and torse, VI, 355, 583—5, 601.

Spinode Curves: and cubic surfaces, VI, 450, 595.

Spottiswoode, W.: ineunts, IV, 419; determinants, IV, 608; text-book on determinants, IV, 608; intersection of line and conic, V, 500—4; twenty-one coordinates of conic in space, XI, 82—3; death of, XI, 430; Sylvester's researches, XIII, 44.

Square Diagrams: seminvariants and solution by, XIII, 288—98.

Squares: products of sums of, II, 49—52; surfaces divisible by curves of curvature into, VIII, 97—8, 145—6, 264—8; theorems of 2-, 4-, 8-, 16-, XI, 294—313; imaginaries of 8-, XI, 368—71; Mill on, XI, 432—3; Latin-, XIII, 55—7; orthomorphosis of, into circles, XIII, 191—202.

Squares of Roots: equations of, IV, 242—3.

Squarewise Contractible: the term, XIII, 179.

Stader, F.: central forces problem, IV, 520, 592.

Standard Solutions: of system of linear equations, XII, 19—21.

Statics: six coordinates of a line, VII, 89—95; and time, XI, 444; and Archimedes, XI, 446.

Stativity: the term, VIII, 213.

Staudt, K. G. C. v.: distances of points, I, 581; theory of distance, II, 605; theorem of, on Bernoulli's numbers, IX, 261.

Steiner, J.: Pascal's theorem, I, 322—8; geometry of position, I, 356, 550—6; transformation of curves, I, 474; conics inscribed in a quadric surface, I, 557; cubic surfaces and triple tangent planes, I, 589; extension of Malfatti's problem, II, 57—86, 593; porism formula, II, 90; harmonic relation of two lines or points, II, 96; in-and-circumscribed polygon, II, 141; problems on quadrics, II, 179—80; pippian, II, 381, 391; systems of circles, III, 113; double tangents, IV, 187; conics, IV, 207; point of six-pointic contact on cubic, IV, 207; conics touching curves, V, 31—2; generalized theorem, V, 100—2; pedals, V, 113; quartic surface, V, 421—3; theorem of eight points on a conic, V, 427—30; Casey's equation, VI, 67; locus, envelope, and triangle, VI, 72; bitangents of quartic curve, VII, 124; quartic surface of, VII, 247, VIII, 389, IX, 1—2, X, 607; attraction of ellipsoidal shell, IX, 302.

Stellated Polyhedra: IV, 82, 609.

Stereographic Projection: of spherical conic, V, 106—9; geodesic lines on oblate spheroid, VII, 24—5; property of, VII, 397—9, XI, 187—9, 569.

Stereoscope: and cubic curves, IV, 122.

Stern, M. A.: sums of certain series, III, 126.

Stirling, J.: curve classification, V, 354; proof of his theorem, X, 267—8.

Stockwell, J. N.: determinants, V, 45—9.

Stokes, Sir G. G.: surfaces, envelopes, and parallel curves, IV, 123—33; resisting medium, IV, 541; Catalogue of Scientific Memoirs, V, 546—8, 620; report on mathematical tables, IX, 461—99.

Streblosis: XI, 79, 81.

Striction, Curve of: the term, I, 234.

Striction, Lines of: on skew surfaces, XIII, 232—7.

Stroh, E: perpetuants, XIII, 266, 301—6, 314—8.

Study, E.: Zeuthen on, VI, 594.

Sturm, J. C. F.: integration of dynamical equations, III, 186, 203.

Sturm, R.: homography, VIII, 200; correspondence of points and tetrahedra, VIII, 200—8; root-limitation, IX, 39; theory of equations, X, 4—5, XI, 498—9, 505.

Sturmian Constants: for cubic and quartic equations, IV, 473—7; for quantics, VI, 159.

Sturmian Functions: note on, I, 306—8, 588; new researches, I, 392—6; endoscopic and exoscopic expressions, I, 588; tables for equations from second to fifth degrees, II, 471—4.

Sturm's Theorem: VI, 159, 161; Sylvester's work at, XIII, 46.

Subinvariants: the term, XII, 251, 273.

Subrational: the term, IX, 315.

Sub-regular Integrals: of differential equations, XII, 444—52.

Substitutions: theorem relative to theory, VII, 47; arising from a problem in arrangements, X, 247—8; and theory of groups, X, 324—30, 401—6; and permutations, X, 574; evolution, XI, 455; the notion of, XI, 509—10, 521; Latin squares, XIII, 55—7; groups for two to eight letters, XIII, 117—49; Sylow's theorems on groups, XIII, 530—3; sixty icosahedral, XIII, 552—7.

Subsurface: the term, IX, 79.

Summit: defined, V, 63, XIII, 507.

Sums: of squares, II, 49—52; the term, X, 186, 192; of two series, XIII, 49—50.

Sun: and moon's mean motion, III, 522—61; Newcomb on parallax, IX, 177—8; (*see also* Solar Eclipse).

Supercurve: the term, IX, 79.

Superlines: in hyperspace, IX, 79—83.

Supp: the term, VI, 263.

Supplement: the term, VI, 263.

Suremain-de-Missery, A.: imaginaries, XII, 467.

Surface, Congruence, Complex, in *Ency. Brit.*: XI, 628—39; introductory, XI, 628—9; surfaces in general: torses, XI, 629—32; surfaces of orders 2, 3, and 4, XI, 632—4; congruences and complexes, XI, 634—5; curves of curvature: asymptotic lines, XI, 635—6; geodesic lines, XI, 636—7; curvilinear coordinates, XI, 637; orthotomic surfaces: parallel surfaces, XI, 637—8; minimal surface, XI, 638—9.

Surface-integral: prepotential, IX, 321—30.

Surface of Centres: for wave-surface, XIII, 248; (*see also* Ellipsoid, Centro-surface of).

Surface of Cylinder: Archimedes' theorem for, XII, 56—7.

Surface of Revolution: and Mercator's projection, VIII, 567.

Surfaces: equimomental, I, 253—4; wave (tetrahedroid), I, 302—5, 587; confocal, I, 362—3; singularities, II, 28—32, IV, 22—7; theory of skew, II, 33—4; envelopes and parallel curves, IV, 123—33, 152—7, 158—65; curvature of, IV, 466—9; theorem on degenerate, V, 98—9; developable, and prohessians, V, 267—83; planar, V, 578; sibi-reciprocal, VI, 21, X, 252—5; sextic, VI, 87—100; singularity of, VI, 123—8; tetrahedral, VII, 48—53; on certain skew, VII, 54—65; Steiner's, VII, 247; intersection of two, VII, 563; divisible into squares by curves of curvature, VIII, 97—8, 145—6, 264—8; correspondence, transformation, and deficiency, VIII, 200—8; penultimate forms of, VIII, 262—3; transformation of unicursal, VIII, 388—93; deficiency of certain, VIII, 394—7; reciprocal, VIII, 394; of eighth order, VIII, 401—3; representation on plane, VIII, 538—9; families of, VIII, 567; transformation of equation of, to chief axes, IX, 48—51; the term, in five-dimensional geometry, IX, 79; orthogonal to set of lines, IX, 587—91; flexure of spherical, X, 30—2; of minimum area, X, 63—7, XIII, 41—2; octic, X, 79—92; on a sibi-reciprocal (octic), X, 252—5; fleflecnodal planes, X, 262—4; Jacobian of six points, X, 281—93; flexure of, X, 331—2; distribution of electricity on two spherical, XI, 1—6; general theory, XI, 14—6, 224; deformation and flexure of, XI, 66—7, 317—22; theory of apsidal, XI, 111—3; theory of reciprocal, XI, 225—34; contact of line with, XI, 281—93; geodesic curvature of

curve on a, XI, 323—30; Gaussian theory, XI, 331—6; and solid geometry, XI, 569; ruled, in *Ency.*
Brit., XI, 572—3; in *Ency. Brit.*, XI, 580—2; general theory of curvilinear coordinates, XII, 1—18;
determination of order of surface, XII, 42—4; minimal, and Joachimsthal's theorem, XII, 594—5;
with plane or spherical curves of curvature, XII, 601—38; quasi-minimal, XIII, 42; the absolute,
XIII, 42; applicable to each other, XIII, 253—64; and systems of tetrads of circles, XIII, 425—9;
of order *n* which pass through given cubic curve, XIII, 534—5; (*see also* Developables, Monoid,
Orthogonal, Parallel, Reciprocal, and Wave Surfaces, Scrolls).

Switzerland: Cayley's visits to, VIII, xi.

Syllogism: theory of, VIII, 65—6.

Sylow, L.: theorems on groups, XIII, 530—3.

Sylvester, J. J.: special factors, I, 337; Sturmian functions, I, 392, II, 471—4; schoolgirl problem,
I, 483; theory of hyperdeterminants, I, 577, 589; commutants, I, 584; endoscopic, I, 588; theory
of permutants, II, 23, 26, 27; rationalization of algebraical equations, II, 42; matrices, II, 219,
604; law of reciprocity, II, 232, 234; partitions, II, 248—9, 506, XII, 217; contravariants, II, 320;
combinants, II, 322; cubic curves, II, 405; symmetric functions, II, 465; canonical forms, II, 523;
cobezoutiants, II, 524; bezoutiants, II, 526; hyperdeterminants, II, 598—601; logic of characteristics,
III, 52; a special determinant, III, 122; elimination, III, 214—5; independent variables in differential
calculus, III, 246; reversion of series, IV, 36, 37, 54—9; canonical form of binary quantics, IV,
43—52, 53; double partitions, IV, 166—70; conics and five-pointic contact, IV, 231; on derivative
of point on cubic, IV, 231; finite differences, IV, 263; equation of differences, IV, 277; invariants,
IV, 349; Tschirnhausen's transformation, IV, 391; volume of tetrahedron, IV, 462; involution of
six lines, IV, 582, 593, VII, 66; lines in involution, V, 1—3; quadric cones, V, 6; quartic surfaces,
V, 69; canonic root of binary quantic, V, 103—5; discriminant of quintic, V, 592; conic and cubic,
V, 608; derivation of points of cubic curve, VI, 20; quintics, VI, 147—8; on roots of algebraical
equation, VI, 147; bicorn, VI, 158; foci of conics, VII, 1—4; differential operators, VII, 8; cubic
transformation of elliptic functions, VII, 44; Cartesian curves and cubic curve, VII, 556; spherical
problem, VII, 563; discussions with, on covariants, VIII, xv; theory of matrices, VIII, xxxii—iii;
root-limitation, IX, 22, 39; elimination, IX, 43, XIII, 545—7; residuation, IX, 211; quartic curves
and functions of a single parameter, IX, 315—7; scalene transformation, IX, 527, 534; development
of idea of Eisenstein, X, 58—9; numerical generating function, X, 339; linkwork, X, 407; N. G. F.
of binary septic, X, 408—9; theorem relating to covariants, X, 430; on trees, X, 598—600; theory
of tamisage, XI, 409—10; partitions, XII, 217; perpetuants, XII, 251, 252, 253; non-unitary partition
tables, XII, 273; d'Alembert-Carnot geometrical paradox, XII, 305—6; umbræ, XII, 347; invariants
and reciprocants, XII, 393; a Diophantine relation, XII, 596; *Nature*, notice in, XIII, 43—8; syzygetic
relations, XIII, 224; reciprocants, XIII, 333—5, 366, 379—81; lectures on theory of reciprocants,
XIII, 381—98.

Symbolical Forms: of hyperdeterminants, I, 80—94; of covariants, I, 577, 585.

Symbols: modular functions, IV, 484—9.

Symmetric: the term, I, 410.

Symmetrical: the term, IV, 599, 604, VI, 524—5, XI, 496.

Symmetric Covariants: II, 233.

Symmetric Curve: and system of equations, I, 473.

Symmetric Functions: of roots of an equation, II, 417—39, 602—3; partitions, II, 418; tables, II, 423—39;
resultant of a system of two equations, II, 440—53; tables, II, 445—53, VI, 292—9; of the roots of
certain systems of two equations, II, 454—64, VI, 292—9; conditions for existence of given systems of
equalities among roots of an equation, II, 465—70, 603—4, VI, 300—12; tables, II, 467; conditions
for existence of systems of equal roots of binary quartic or quintic, VI, 300—12; and theory of
equations, X, 6—8; non-unitary, and seminvariants, XII, 239—48, 275; tables of roots, XII, 263—72,
273—4; a differential operator, XII, 318; seminvariants, XIII, 265—332; (*see also* Seminvariants).

Symmetroid: the term, VII, 134, 280; lineo-linear correspondence of quartic surfaces, VII, 157—9; and Jacobian, VII, 160—3, 175; with given nodes, VII, 163—6; and decadianome, VII, 256, 259; and circumscribed cone, VII, 258—9; theory, VII, 264.

Symmetry: Sylvester on, XIII, 45.

Symptose: the term, I, 523, 529, 557—8.

Syntypic: the term, VII, 123.

Système Linéaire: of Laguerre is a matrix, II, 604.

System of Equations: order of, I, 457—61, 589; connected with Malfatti's problem, I, 465—70; note, I, 532—3, 589; algebraical, XI, 39—40.

Syzygant: the term, XII, 251; and seminvariants, XII, 257—62.

Syzygies: of degree six, VI, 148—53; of binary quintic connected, VII, 334; for binary cubic, IX, 55; of quintic, X, 346—55; of sextic, XII, 257—62; of binary quartic, and elliptic integrals, XIII, 32; Sylvester's work in, XIII, 46; syzygetic relations among powers of linear quantics, XIII, 224—7; and seminvariants, XIII, 310.

Tables, Brit. Assoc. Report on Mathematical: IX, 461—99; introductory, IX, 461—2; of divisors and prime numbers, IX, 462—70; prime roots, IX, 471—7; Pellian equation, IX, 477—80; partitions, IX, 480—3; quadratic forms, IX, 484—6; binary, ternary, quadratic, and higher forms, IX, 486—93; complex theories, IX, 493—9.

Tables: linear transformations, I, 108; of covariants for quadratic, cubic, quartic, quintic, II, 276—81, II, 310—35; of covariants M to W of binary quintic, II, 282—309; covariants for sextic, II, 314—5; for septimic, II, 315—6; for octavic, II, 316—8; for nonic, II, 318—9; of concomitants of ternary quadric, II, 322—3; of ternary cubic, II, 323—9, 331—5; of symmetric functions of roots of equation, II, 423—39; of resultants of two equations, II, 449—53; Sturmian functions for equations from second to fifth degrees, II, 471—4; disturbing function in lunar theory, III, 299—308, 311—8, VII, 516, 519—24, 525—7; of functions in theory of elliptic motion, III, 360—474; Degen's, for Pellian equation, IV, 40; equation of differences, IV, 246—56, 280—91; Arbogast's method of derivations, IV, 274—5; concomitants of ternary cubics, IV, 333—41; Tschirnhausen's transformation for quartics, IV, 373—4, 379—80; and for quintics, IV, 387—90; numerical expansions, IV, 470; polyacra, V, 44; binary quadratic forms, V, 141—56, 618; properties of scrolls, V, 171—2; axial systems of polyhedra, V, 532—9; curves in space, V, 616; for prime or composite modulus, VI, 83—6; asyzygetic covariants, VI, 149—152; quantics, VI, 167—8; resultant of a system of two equations, VI, 292—9; conditions for existence of systems of equal roots of quartic or quintic, VI, 300—12; singularities of cubic surfaces, VI, 363; also lines and planes, VI, 373; Legendre's elliptic functions, VII, 20; geodesic lines on oblate spheroid, VII, 23; rational transformation between two spaces, VII, 210—3, 224; nodal quartic surfaces, VII, 283, 287, 291, 296; quartic surfaces, VII, 310, 609—10; irreducible covariants of binary quintic, VII, 341—6; planogram No. 1, VII, 439—40; ditto No. 2, VII, 450—1; geodesic lines on ellipsoid, VII, 504—6; binary cubic forms, VIII, 51—64; theory of curve and torse, VIII, 81—4; Pineto's of logarithms (review), VIII, 95—6; cones satisfying six conditions, VIII, 100; geodesic lines, particularly on quadric surface, VIII, 196—9; in-and-circumscribed triangle, VIII, 214—21; centro-surface of ellipsoid, VIII, 365; Steiner's surface, IX, 7; transformation of elliptic functions, IX, 128—35, 163; Newcomb's planetary, IX, 181—4; projection of skew hyperboloid of revolution, IX, 240; classification for mathematical, IX, 424—5; report on mathematical, IX, 424—5; chemical trees, IX, 436—43, 446—8, 450—60, 544—5; double theta functions, X, 168—9, 171, 172—3; regular solids, X, 270—3; concomitants of quintic, X, 349—55, 362—9, 370—6, 377—94, 397—400; transvectants for quintic, X, 378—394; Kummer hexads, X, 506; theta functions, X, 507—10, 513—28, 530—6, 540—2, 544—6; theory of numbers, trisection, XI, 89; ditto quartisection, XI, 94; Reuschle's, of prime roots, XI, 95—6; of finite differences, XI, 144—7; connected with polyhedral function, XI, 158—9, 192; covariantive, XI, 272—80; Schubert's numerative geometry, XI, 286; theorems of squares, XI, 299—313; theory of numbers, XI, 316; concomitants of ternary cubic, XI, 345—7; literal, for binary quantics, otherwise a partition table, XI, 357—64; for binary sextic, XI, 372—6, 377—88; covariantive, XI, 409—10; Plücker's equations, XI, 472; of Gauss, XI, 545; symmetric functions of roots of an equation, XII, 263—72, XIII, 272—4, 288; non-unitary partition, XII, 273—4; seminvariant, XII, 275—89; orthomorphosis of circle into parabola, XII, 336; of groups, orders two to twelve, XII, 643—56; Wallis's investigation for π, XIII, 23—5; quadrinvariant and cubinvariant of quadri-quadric function, XIII, 68; partitions of a polygon, XIII, 95, 112; theory of rational transformation, XIII, 116; substitution groups for two to eight letters, XIII, 118—49; corrected seminvariant for weights 11 and 12, XIII, 217—23; of conjugates, XIII, 303; seminvariants and symmetric functions, XIII, 311, 313, 331—2; of pure reciprocants to weight 8, XIII, 333—5; report of British Association committee on Pellian equation, and tables, XIII, 430—67; omega and theta functions, XIII, 558—9.

Tacinvariant: the term, IV, 607.

Tac-locus: in singular solutions, VIII, 533.

Tacnode: defined, II, 28—32, V, 266.

Tactic: and algebra, V, 293—4; the term, XI, 443; (*see also* Arrangements, Groups).

Tactinvariant: of two quantics, II, 320; the term, V, 305.

Tactions: analytical solution, III, 255—7; formulæ, IV, 510—2; Casey's equation, VI, 543; and trizomal curves, VI, 575; problem of, XIII, 150—69.

Tait, P. G.: arrangements, X, 245; quaternions, XII, 303, 475, XIII, 541—4; finite differences, XII, 412.

Talbot, W. H. F.: curve of, IV, 123.

Tamisage: Sylvester's theory of, XI, 409—10.

Tangential: defined, II, 558; of a curve, IV, 188.

Tangent Lines: and surface, XI, 630, 632—4.

Tangent Omals: VI, 467—9.

Tangent Planes: and surface, XI, 630, 632—4.

Tangents: and two-dimensional geometry, II, 575; inflexional and chief, VIII, 157, 294; singular of a quartic, X, 603; in *Ency. Brit.*, XI, 564—5, 579—80; (*see also* Bitangents).

Tannery, J.: linear differential equations, XII, 394.

Tantipartite: the term, I, 584, II, 517, IV, 464, 604.

Taylor, C.: locus *in plano* problem, VII, 599; general theory of surfaces, XII, 42—4.

Taylor, H. M.: inversion, IX, 18; partitions of a polygon, XIII, 93, 112.

Taylor's Theorem: Lagrange's demonstration, VIII, 493—5, 519; note on, VIII, 524.

Tchebycheff, P.: theory of numbers, XI, 616.

Terminology: recent mathematical, IV, 594—608.

Terms: in symmetrical determinant, IX, 185—90.

Ternary: the term, IV, 604, VI, 464.

Ternary Cubics: relation between two, IV, 79—81; memoir on quantics, IV, 325—41; form problem, VII, 548; the 34 concomitants, XI, 342—56; canonical form, XI, 343.

Ternary Quadratics: resultant of three, IV, 349—58.

Ternary Quadrics: and involution, XIII, 350—3.

Ternary Quantics: and bitangents of plane curve, IV, 188; involution, V, 301—9.

Tetrads: the term, XII, 590; systems of, XIII, 425—9.

Tetrahedra: reciprocals, III, 7; axial systems, V, 531—9; note on, V, 557—9; and cubic surfaces, VII, 607; correspondence of points in relation to two, VIII, 200—8; Steiner's surface, IX, 1—12; in perspective, IX, 209—10; automorphic function for, XI, 169, 179—83, 184, 212—6.

Tetrahedral Surfaces: VII, 48—53, 54—65.

Tetrahedroid: and wave surface, I, 302—5, 587; 16-nodal quartic surfaces, V, 431—7; the term, VI, 21, X, 252; and scrolls, VII, 245; as particular case of 16-nodal quartic surface, X, 437—40.

Tetrazomal (*see* Polyzomal Curves).

Text-Books: on determinants, elimination and higher algebra, IV, 608.

Theory of Equations (*see* Equations, Theory of).

Theory of Groups (*see* Groups).

Theory of Numbers (*see* Numbers, Theory of).

Theta Functions, Memoir on Single and Double: X, 463—565; historical, X, 463—4; Part I, X, 464—76; definitions, X, 464—5; allied functions, X, 465—6; even-integer alteration of characters, X, 466; odd ditto, X, 466; even and odd functions, X, 467; quarter-periods unity, X, 467—8; conjoint quarter quasi-periods, X, 468—9; product-theorem, X, 469—71; résumé of ulterior theory of the single functions, X, 471—3; ditto, double functions, X, 474—5; remark as to notation, X, 475—6; Part II, X, 476—97; notation, X, 476; constants of the theory, X, 477—8; product theorem, X, 478—80; the square-set, X, 481—2; relation between the constants, X, 482—3; product-sets, X, 483—4; comparison with Jacobi's formulæ, X, 485; the square set, X, 485—7, 488—9; elliptic integrals of third kind, X, 489—90; addition formulæ, X, 491—2; doubly infinite product forms, X, 492—4; transformation q to r, X, 494—7; Part III, the double theta functions, X, 497—565; product-theorem, X, 497—506; tables, X, 506—8; product-theorem and its results, X, 509—39; tables, X, 513—39; the first set, X, 539; second ditto, X, 540; third ditto, X, 541; fourth ditto, X, 542; considerations, X, 543—8; résumé, X, 548; 16-nodal quartic surfaces, X, 548—51; x, y expressions of theta functions, X, 551—5; further results of product-theorem, X, 555—7; differential relations connecting theta and quotient functions, X, 557—9; differential relations of theta functions, X, 559—61; ditto, u, v, x, y, X, 561—5.

Theta Functions: of Jacobi, I, 136, 290; and elliptic integrals, XI, 41—6; theory of multiple, XI, 242—9; notation, XI, 243—5; evolution, XI, 451—5; the term, XI, 532; linear transformation, XII, 337—43; formula relating to zero value of, XII, 442—3; Smith's memoir, XIII, 558—9; (*see also* Abelian, Double Theta, Elliptic, Single Theta, and Triple Theta, Functions).

Third Class: curves of, II, 395—6.

Thomae, J.: linear differential equations, XII, 394, 396, 444; theta functions, XII, 442.

Thomson, F. D.: tangents of conic, V, 578.

Thomson, J.: mechanical integrator, XI, 53.

Thomson, W. (*see* Kelvin, Lord).

Three-bar Motion: IX, 551—80, XI, 481, XIII, 505—16.

Three Bodies: problem of, III, 97—103, 183, IV, 548—552; in a line, IV, 538—40; other cases, IV, 540—1.

Time and Number: XI, 442—4.

Tissot, A.: spherical pendulum, IV, 534, 593.

Titus, Colonel: arithmetical problem, IV, 171—2.

Todhunter, I.: conics, IV, 481; Taylor's theorem, VIII, 493—5; q-squares, X, 27; probabilities, X, 600.

Topography: contour and slope lines, IV, 108—11, 609.

Topology: of space, VI, 22; of chessboard, X, 609.

Torsal: the term, VI, 334, 336, 341, 355, 582—5.

Torse, on a Certain Sextic: VII, 99—114; introductory, VII, 99—100; theorem of four binary quartics, VII, 100; standard equation of unicursal quartic, VII, 101; tangent line and osculating plane of unicursal quartic, VII, 101; its final form, VII, 102; determination of sextic torse, VII, 102—3; principal sections of ditto, VII, 103—5; partial determination of equation, VII, 105; determination of the unknown coefficients, VII, 106—11; equation of sextic torse, VII, 112; ditto, and centro-surface of ellipsoid, VII, 113—4.

Torses: the term, V, 182, XI, 573; and scrolls, V, 199—200; and curves, V, 505—10; a special sextic developable, V, 511—9; singularities, VI, 601; on some sextic, VII, 116—7, 118—20; circumscribed to two quadrics, VIII, 520—1; on a sextic, X, 68—72; depending on elliptic functions, X, 73—8;

and certain octic surfaces, x, 79—92; kinds of, xi, 227; in *Ency. Brit.*, xi, 628, 629—32; and surfaces, xi, 632; and non-Euclidian plane geometry, xii, 222; (*see also* Developables).

Torsion: the term, i, 234, xiii, 232, 234.

Tortolini, B.: envelopes, parallel curves and surfaces, iv, 123—33; parallel surfaces of ellipsoid, iv, 133.

Tortuous Curves (*see* Curves).

Torus: the term, vii, 246, viii, 25; paper by Darboux, vii, 247; the conic-, ix, 519—21.

Townsend, R.: inertia, iv, 566, 593; confocal quadrics, viii, 520.

Tractor: the term, vii, 73—5, x, 269; six coordinates of a line, vii, 85—6, 93—5.

Trajectories: root-limitation, ix, 22—7; and orthomorphosis, xiii, 170.

Transcendental Analysis (*see* Function).

Transcendental Function: the term, xi, 524.

Transcendental Integrals (*see* Abelian Integrals).

Transcendent, Gudermannian: v, 86—8, 617.

Transformation: of quadratic forms, ii, 145—9, 192—201, 215; of two quadric functions, iii, 129—31; the term modulus of, iv, 605; plane curves, vi, 1—8, 593, viii, 387; Cremona's, vi, 22—3; polyzomal curves, vi, 553, 565—6; two quantics into each other, viii, 385—7; unicursal surfaces, viii, 388—93; binary quadratic form, viii, 398—400; doubly infinite products, x, 494—7; theories, xi, 482; Landen's, xi, 584; double theta functions, xii, 358—89; of order 11, and modular equation, xiii, 38—40; modular equation for cubic, xiii, 64—5; (*see also* Special Headings below).

Transformation, Automorphic: iv, 416, v, 439; of binary cubic function, xi, 411—6.

Transformation, Cubic: in elliptic functions, ix, 522—6.

Transformation, Geometric: vii, 121—2.

Transformation, Homographic: xi, 189—90, 196—208.

Transformation, Linear: ii, 225, xi, 237—41; imaginary linear, vi, 183—6; lineo-linear, vii, 215—6, 236—8; of theta functions, xii, 337—43.

Transformation of Coordinates: i, 123—6, 586, iv, 552—9, vii, 95, 415—7, xi, 136—42, 558—61; formulæ, vii, 97—8.

Transformation of Elliptic Functions: i, 120—2, 585, v, 472, ix, 103—6, 244—5, x, 333—8, 611, xi, 26, xii, 416—7, 535—54, xiii, 29—32, 490—2, 505—34, 535—55, 556—7.

Transformation of Elliptic Integrals: i, 508—10, iv, 60—9, 609.

Transformation of Equations: ix, 42, 48—51; of differential, v, 78—9.

Transformation of Integrals: i, 383, iii, 1—4, 438—44, ix, 250—2.

Transformation of Tschirnhausen: vi, 165—9, xi, 396; for cubics, iv, 364—7, xiii, 421; quartics, iv, 368—74; quartics and quintics, iv, 375—94, v, 449—53; theory of equations, xi, 509.

Transformation, Orthomorphic: of a circle into itself, xiii, 20.

Transformation, Quadric: of elliptic functions, xii, 58; between two planes, xii, 100—1.

Transformation, Rational, between Two Spaces, Memoir: vii, 189—240; introductory, vii, 189—90; general principle, vii, 190—3; homographic transformation between two lines, vii, 193—7; rational ditto between two planes, vii, 197—213, 216—21; tables, vii, 210—3; quadric transformation between two planes, vii, 213—6; quadric transformation any number of times repeated, vii, 219—21; reduction of general rational transformation to a series of quadric transformations, vii, 222—4; rational transformation between two spaces, vii, 224—9, 238—40; ditto quadri-quadric, vii, 229—30; ditto quadri-cubic, vii, 230—3; ditto cubo-cubic, vii, 233—8; this principal system consists of six lines, vii, 234—6; principal system of a proper sextic curve—the lineo-linear transformation between two spaces, vii, 236—8.

Transformation, Rational: of plane curves, vi, 1—8; does not alter deficiency, vi, 3; between two planes and special systems of points, vii, 253—5; note on a theory of, xiii, 115—6.

Transformation, Rectangular: xi, 421—8.

Transformation, Scalene: of plane curve, ix, 527—34.

Transformation, Septic: of elliptic functions, x, 333—8, xII, 535—54.

Transformation, Special Quartic: of elliptic functions, IX, 103—6.

Transmutant: defined, II, 515.

Transpose: the term, II, 493.

Transvectant: form of covariants, VIII, 404—8; (*see also* Derivatives).

Trees: analytical forms called, III, 242—6, IV, 112—5, XI, 365—7; curves which satisfy given conditions, VI, 260; application to chemistry, IX, 202—4, 427—60, 544—5; problem and solution, x, 598—600; a theorem on, XIII, 26—8.

Triads: of seven and fifteen things, I, 481—4, 589; of fifteen things, v, 95—7.

Triangle: harmonic relation of point and line, II, 96—7; reciprocal triangles, III, 5—7; circumscribed about conic, properties of, III, 29—34; theorem of line and conic, v, 100—2; problems, v, 564, 566, 593, VII, 581, 599, x, 575; locus in relation to, VI, 53—64; locus and envelope, VI, 72—82; solution of problem in *Principia*, Bk I. Sec. V. Lemma XXVII, VII, 30; potential of, IX, 270—1; non-Euclidian, XIII, 482—3; nine-point circle of a plane, XIII, 520—1.

Triangle, In-and-circumscribed, the Problem of, Memoir: VIII, 212—57; introductory, VIII, 212—3; tables, VIII, 214—21; principle of correspondence, VIII, 222—5; locus of a free angle, VIII, 225—7; application of theory to locus, VIII, 227—8; solutions for 52 cases, VIII, 228—51; the case 52, VIII, 251—57.

Triangle, In-and-circumscribed: II, 87—90, 91—2, 138—44, 145—9, III, 67—75, 229—41, IV, 435—41, v, 489—92, 549—50, 553, VIII, 565—6; a posteriori demonstration of porism, III, 80—5; (*see also* Porism).

Triangle, Spherical: theorem of, IV, 80, XI, 97—9; nine-point circle of, XIII, 548—51.

Tricircular Sextic: IX, 562—70.

Trident Curve: classification, v, 355—69, 395.

Trigonometry: transformation of an expression, II, 45—6; multiple sines, x, 1—2; theorem in partitions and, x, 16; identities, XI, 38, XIII, 538—40; formulæ, XII, 108; an expansion, XII, 319—20.

Trigonometry, Spherical: theorem, IV, 80, XI, 97—9; identity, VIII, 525; foundation, XI, 570.

Trigonometry, Spheroidal: IX, 197.

Trihedral Pair: the term, VI, 374.

Trinodal Quartic: x, 602; (*see also* Quartic Curves).

Tripair: the term, x, 450—1.

Tripartite: the term, VI, 464.

Triple Tangent Planes: of cubic surfaces, I, 445—56, 589, VI, 372, 376.

Triple Theta Functions: x, 432—6, XI, 47—9; algorithm for characteristics of, x, 441—5; and quartic curves, x, 446—54.

Tris: the abbreviation in groups, XIII, 119.

Trisection: in theory of numbers, XI, 84—96.

Tritom: and point, v, 138.

Trivector: the term, VII, 400, 401; planet's orbit from, VII, 406—12, 426—8.

Trizomal (*see* Polyzomal Curves).

Trope: the term, VI, 330, VIII, 73, x, 54—5.

Tropical Point: the term, XII, 433.

Truel, H. D.: imaginaries, XII, 467.

Tschirnhausen's Transformation: VI, 165—9, XI, 396; for cubics, IV, 364—7, XIII, 421; quartics, IV, 368—74, 375—82, v, 449—53; quintics, IV, 382—94; theory of equations, XI, 509.

Tucker, R.: geometrical interpretation, x, 581.

Twisted: the term, VI, 524—5.

Two Centres Problem: IV, 524—32.

Two-way Point: the term, XIII, 507.

Ueberschiebung: the word, I, 585; (*see also* Derivations).

Ultra-elliptic Functions (*see* Hyperelliptic Functions).

Umbilicar Centres: the term, VIII, 326, 351.

Umbilici: and differential equations, V, 115—30; curves of curvature near, VII, 330—1; on surface of nth order, VIII, 320; the term, XI, 581; (*see also* Geodesic Lines).

Umbral: theory of seminvariants, XIII, 266; notation of, and seminvariants, XIII, 301—6.

Umbræ: the term, XII, 347.

Unibasic: the term, XII, 642.

Unicursal Curves: VI, 2.

Unicursal Octics: XII, 310.

Unicursal Quartics: standard equation, VII, 101; tangent line and osculating plane, VII, 101; its final form, VII, 102.

Unicursal Surfaces: transformation of, VIII, 388—93.

Unicursal Twisted Quartic: XII, 428—31.

Uniform Convergence: XIII, 342—5.

Uniform Function: XII, 433.

Uniform Series: defined, IV, 457.

Unipartite: the term, VI, 464.

Uniplanar-node: the term, VI, 361.

United Points: in correspondence, VI, 9.

Unity: prime roots, XI, 56—60; imaginary roots, IX, 263; ninth roots, XIII, 66.

Universal Algebra: Sylvester's theory of, XIII, 47.

Unode: the term, VI, 362.

Uranus: Newcomb's work, IX, 180—4.

Vacuity: Sylvester's theory of, XIII, 47.

Valentiner, H.: curves in space, V, 613—7; theory of surfaces, XI, 14—6.

Values: principal, of complex expression, I, 309; of $\Pi i = \Gamma (1 + i)$, XIII, 522—4.

Vandermonde, A. T.: solution by radicals, X, 11; theory of equations, XI, 513.

Variables: (2, 2) correspondence of two, IX, 94—5; normal in dynamics, IX, 111; imaginary, XI, 439—41.

Variation: of parameters in rotation of solid body, I, 242; of arbitrary constants, III, 161—200; of planet's orbit, III, 516—8, VII, 541—5.

Variations, Calculus of: Jacobi on, III, 174; problem in, VII, 263.

Velocities, Virtual: general equation, IX, 205—8.

Veronese, G.: Pascal's theorem, VI, 594; four-dimensional space, XI, 442.

Vertices of Cones (*see* Cones).

Vicinal Surfaces: conormal correspondence of, VIII, 301—8; (*see also* Surfaces).

Vieta, F.: tactions, XIII, 152.

Virginia: Key and Sylvester, professors at, XIII, 43.

Virtual Velocities: general equation, IX, 205—8.

Walker, S. C.: the anharmonic-ratio sextic, VII, 314—5; orbit of Neptune, IX, 180.

Wallis, John: biographical notice, XI, 640—3; multiple algebra, XII, 466; his expression for π, XIII, 22—5.

Walton, W.: root-limitation, IX, 39; maxima and minima, IX, 40—1; transformation of equations, IX, 42; integration and definite integrals, IX, 56—63; ray planes and biaxal crystals, IX, 107—9.

Waring, E.: equation of differences, IV, 240, 252; sum of mth powers of the roots of an equation, XIII, 213—6.

Warren, J.: on binary cubics V, 289; curvilinear coordinates, XII, 1—18; multiple algebra, XII, 460, 468.

Wave Surfaces: tetrahedroid, I, 302—5, 587, VI, 21; on, IV, 420—6, 432—4, XIII, 238—52; equation of, in elliptic coordinates, XI, 71—2; evolution of Fresnel's, XI, 449.

Weber, H. H.: triple theta functions, X, 444, 446—54; bitangents of quartic, XI, 221—3; elliptic functions, XIII, 559.

Weddle, M.: cubic curves, IV, 497; quadric cones, V, 4.

Weierstrass, K.: doubly infinite product, I, 586; function $Al(x)$, I, 587; al-functions, V, 33—7; Steiner's quartic surface, V, 423; infinite products, VIII, xl; triple theta functions, X, 432, 434; theta functions, X, 499; elliptic integrals, XI, 64; theta functions, XI, 242; theory of functions, XI, 451—2, 454; function of, XI, 540; Abelian functions, XII, 98; and Jacobian elliptic functions, XII, 425—7; transformation in elliptic functions, XIII, 29, 31.

Weight: and partitions of a polygon, XIII, 110.

Weingarten, J.: application of surfaces to each other, XIII, 253—64.

Whewell, W.: dynamics, IV, 518; mathematics, XI, 431—2; number and time, XI, 442.

Whitworth, W. A.: triangles and conics, V, 593.

Wiener, C.: model of cubic surface with twenty-seven real lines, VIII, 366—84.

Wilbraham, H.: probabilities, II, 594—8, V, 85.

Wilkinson, M. M. U.: Taylor's theorem, VIII, 519; chances, X, 588; rectangular transformation, XI, 421—8.

Wilkinson, T. T.: circle and points, V, 560.

Wilson, J.: theorem of, XI, 598; and proof, XII, 45.

Wolstenholme, J.: relation among derivatives of a function, X, 590—2; conic and cubic, X, 605—7.

Women: Cayley and higher education of, VIII, xix.

Woolhouse, W. S. B.: theorem of integration, problem, VII, 588; algebraical theorem, X, 594—6.

Worms, H.: rotation of the Earth, IV, 537, 593.

Woven: the term, XIII, 121.

Wright, E.: Mercator's projection, XI, 448.

Wright, T. C.: on Cayley as a law-student, VIII, xiv.

Writing of Cayley: frontispiece, VIII.

Wronski, H.: theorem of, IX, 96—102.

Young, G. P.: soluble quintic equations, XIII, 88; theory of groups, XIII, 336, 533.

Young, J. R.: sums of squares, II, 52; theorems of squares, XI, 294, 301.

Zech, P.: wave surface, IV, 420—5, 432—4.

Zero-values: of theta functions, X, 499—500, XII, 442—3; (*see also* Theta Functions).

Zeuthen, H. G.: curves and developables, I, 587; sextactic points, V, 545; curves which satisfy given conditions, VI, 191, 192, 200—26, 594; capitals, VI, 280; reciprocal surfaces, VI, 577—81, 591, 596—601, XI, 234; cubic surfaces, VI, 595—6; correspondence of two points on a curve, VII, 39; theory of conics, VII, 552—4; theory of curve and torse, VIII, 72; table of singularities of torse, VIII, 81—2; degenerate forms of curves, XI, 220; quartic curves, XI, 480; systems of curves, XI, 486—7.

Zolotareff, G.: elliptic integrals, X, 143.

Zomal: defined, VI, 473; (*see also* Polyzomal Curves).

Zornow, A. R.: mathematical tables, IX, 486.

CAMBRIDGE: PRINTED BY J. AND C. F. CLAY, AT THE UNIVERSITY PRESS.

Printed in the United States
By Bookmasters